DR NICKY HAYES is a psychologist and the author of over twenty psychology books. She is an honorary life member of the Association for the Teaching of Psychology, a Fellow of the British Psychological Society and a Council Member of the International Test Commission. She has contributed to numerous academic publications, edited journals and has taught psychology from early years to PhD level. In 1997, she received the British Psychological Society award for Distinguished Contributions to the Teaching of Psychology.

WHAT ARE YOU THINKING?

Why We Feel and Act the Way We Do

NICKY HAYES

Michael O'Mara Books Limited

First published in Great Britain in 2022
by Michael O'Mara Books Limited
9 Lion Yard
Tremadoc Road
London SW4 7NQ

A CIP catalogue record for this book is available from the British Library.

Papers used by Michael O'Mara Books Limited are natural, recyclable products made from wood grown in sustainable forests. The manufacturing processes conform to the environmental regulations of the country of origin.

ISBN: 978-1-78929-380-7 in hardback print format
ISBN: 978-1-78929-381-4 in ebook format

1 2 3 4 5 6 7 8 9 10

www.mombooks.com

Designed and typeset by Design 23

Printed and bound by CPI Group (UK) Ltd, Croydon, CR0 4YY

CONTENTS

INTRODUCTION

Thinking is so much a part of what we do as human beings that we take most of it for granted. But our thinking is more complicated than it appears on the surface, and it has so many facets. From decision-making to passing judgement on ourselves and others, how we think and the way our minds form those thoughts is central to who we are and how we see others. Thinking is so important, in fact, that it isn't just the province of one single area of psychology. Cognitive psychologists, as their name suggests, carry out research into cognition – that is, mental activity – and that includes thinking. A lot of our knowledge about how people think comes directly from their cognition studies – for example, how we represent information in our minds and the various kinds of memory we use.

But we have learned from other areas of psychology too. Neuropsychologists, for example, study how the brain works. They use scanning techniques to observe which parts of the brain are active when we think – such as when we are working out the solution to a puzzle – and that aspect of brain research is known as cognitive neuropsychology. Some social psychologists also specialize in social cognition: those parts of our thinking concerned with how we connect with other people, whether that's by making judgements or in working out how we should act in a social situation. And some facets of our thinking bring together these various aspects of psychology, including the study of empathy or how our biological rhythms affect how we think.

The aim of this book is to explore the main areas of thinking which have the greatest impact on our everyday lives. We'll begin by looking at how we make decisions and solve problems: possibly the first thing that comes to mind when we think about thinking. In the process, we'll learn how we can become trapped into mistakes by taking what seem to be easy options and how we can overcome some of these errors. Our thinking also involves making judgements – about other people and also about ourselves, and this is what we will focus on in the second part of the book. We'll explore, for example, how we habitually judge others differently from the way that we judge ourselves, and how negative explanations can lead to depression. We'll then go on to look at creativity and insight, and consider how much of our other thinking is influenced by our habits and general mental states.

From how we think, we'll move on to the question of how we acquire the information to think with in the first place. There's the fascinating question of how we make sense of data that we're receiving from the outside world: how we recognize things and how we use that information to take action when we need to, and also the way that our perception can be tricked. From there, we'll examine the various techniques we employ to help us represent and store data. But we also need to recall it, so we will look at the many different methods that we use to remember things, ranging from our memory of physical actions and skills to how we use language, and the way that we recall plans and intentions that haven't yet happened. Of course, we don't just remember stuff, we also forget it – and that can happen for many reasons, even unconscious wish-fulfilment. Putting all that together and recognizing that quite a lot of our thinking isn't even conscious, we can appreciate how complicated thinking really is. But it's also fascinating, and knowing about the different facets of thinking and how our minds can be influenced without our realizing it helps us to understand not only ourselves, but also other people.

1

MAKING DECISIONS AND SOLVING PROBLEMS

In this first part, we'll explore some of the ways that we think when we are dealing with problems or challenges in everyday life. A lot of the time we don't really think at all – we just take things for granted and go along with our daily routines. But at other times we concentrate hard – so what makes the difference? And are we still aware of what's around us when we're really concentrating?

Our thinking is full of shortcuts – and these can often lead us into mistakes. Psychologists have found over a hundred different ways that we can be misled by the shortcuts, or heuristics, that we use, ranging from just making the most familiar choices to assuming that our own experience is typical of other people's.

1
CAN YOU WALK AND CHEW GUM AT THE SAME TIME?

Saying someone can't walk and chew gum at the same time is an old insult. It implies that the person has such limited brain power that they have to concentrate really hard on things that other people would do automatically.

Most of our thinking is really so automatic that we barely notice it. We respond to greetings from other people with a routine reply like 'I'm fine, how are you?' without thinking about what we are actually saying. It surprises us if our words are taken literally. We've all been caught out by people who reply to the polite enquiry of 'How are you?' by telling us *exactly* how they are – in great and tedious detail! It certainly teaches you to notice how you respond to such questions. But as a general rule, most people take the words in the spirit they were meant – as a friendly greeting rather than as an invitation to share their entire medical history.

In the same way, we make simple decisions and work out easy problems more or less automatically. They don't interfere with whatever else we are doing and it's how we deal with most daily events and decisions. Nobel Prize-winner Daniel Kahneman called this **system 1 thinking** – an automatic, undemanding type of thinking that allows us to multitask easily. For example: we can mull over the possibilities for tonight's dinner while we're out walking the dog without breaking stride. That's system 1 thinking. It's the same when we decide that we like someone who smiles at us, or work out that 2 + 3 = 5.

System 1 thinking: our everyday thinking, which is fast and intuitive but can be often inaccurate and biased.

But if someone asked us to do a really complicated problem while we were walking the dog – for example, calculating how many days to go before the next 29 February – the chances are that we'd have to stop walking while we worked it out. Some things need concentrated thought, which shuts down other types of mental processing. Kahneman called that **system 2 thinking**: it's the type of logical, systematic thinking that demands our full attention.

Most of the time, Kahneman says, we use system 1 thinking. It's straightforward and doesn't take much cognitive work. But it's also pretty flaky – we can fall into all sorts of traps and errors because it's so habitual. We'll be looking more closely at some of those traps later in this part of the book. In system 1 thinking, we fall back on three things: the everyday assumptions that we make out of habit and don't really think about; the 'received wisdom' that we have simply accepted as being true because that's what we learned from family and elders when we were younger; and

Avoid this patient

One of the examples of fast and slow thinking that Kahneman gives is that of psychotherapy trainees faced with a particular kind of client: someone who arrives for therapy and reveals that there have been a number of failed attempts by other therapists to deal with his problems, before assuring the new therapist that he / she is different from all the others and would understand. The tutor's strongly worded advice was that the trainees shouldn't even think of accepting such a patient. Their first reaction – their system 1 thinking – would be one of sympathy for the patient and a belief that they really could help this time. But they should override that with the rational premise, supported by the evidence (system 2 thinking), that attempting it would be pointless and self-destructive given the patient's past track record. If the person had only seen one or two therapists before, well, maybe. But having seen several then, no, it wasn't at all likely that the trainees could help. Any attempt to do so would be likely to lead them into professional difficulties.

the shared social representations about what causes what and why things are like they are, which are taken for granted in our culture and society. These usually work for routine, undemanding situations. But they don't work well if we really need to work something out clearly and accurately. Then, we think quite differently.

System 2 thinking: logical and deliberate thinking, which is slower and needs concentration but is reasonably accurate.

System 2 thinking is less impulsive and more cautious. It takes up more of our mental energy because we don't just jump to the first available solution. Instead, we look closely at the problem, work out what we need to do and deal carefully with the steps we need to take to work it out – or at least, that's what we try to do. We can still be fooled by attributional errors or mental sets, though, and we'll also explore those later in this book.

When we say that system 2 thinking uses up more mental energy, we mean it. It actually uses physical energy as well, which is why focusing on something for a long time can leave you feeling so tired. But it can be enjoyable too: people do crosswords or Sudoku puzzles for pleasure, and that's using system 2 thinking. And writers, artists and others who concentrate hard for long periods of time often describe how satisfying that process can be. Mental exercise, like physical exercise, can be exhausting, but it can also leave you feeling good.

2
WHAT GORILLA?

Our everyday thinking isn't random. We don't approach any problem or challenge completely naively – even if we are trying to. How we see things is governed by a range of factors: our previous experience, how familiar the situation is, the reason we are addressing the challenge in the first place and, perhaps most of all, our expectations.

We come to any situation with a set of expectations, and that gives us a mental framework for making sense of what we see. It also makes us ready to think in certain ways rather than others. That framework is known as a **mental set**. The word 'set' as psychologists use it doesn't mean a group of things or thoughts. It's used in the same way as someone who starts a race will say 'Get ready – get set – go!' It means being ready, focused or prepared to act (or think) in a particular way.

Mental set: a state of readiness or preparedness in our thinking, which makes us particularly likely to notice a certain type of information.

Mental set is a powerful cognitive mechanism. It affects how we go about solving problems, in that we are more likely to use a familiar strategy, which we know has worked in the past, rather than to try a new one – even if the new one would be more effective. In a classic study showing this process, people were given a series of problems in which they were asked to work out how to measure exact amounts of water using only three jars of different capacity. The first few problems could only be solved by following the same quite complicated sequence of pouring water between the large, medium and small jars. A later problem had a much easier solution, but the participants in the study had become so used to the complicated approach that they couldn't see the simple answer at all. So they solved that problem in the same way that they'd tackled the earlier ones, which was much more complicated than it need have been.

There are more dramatic examples of mental set, though. One amazing illustration of how much it can affect our cognition came in a classic study by the psychologists Christopher Chabris and Daniel Simons. They asked people to watch a video of a basketball game. One team's players wore white shirts and the other side wore black. The viewers were asked to count how many times members of the white-shirted team passed the ball, but to ignore the passes made by the black-shirted players. It was a fast game, so the task needed concentration. During the footage, a woman in a gorilla suit strode right across the front of the screen, stopping dead centre, then beat her chest before walking off. The gorilla was in plain view for about nine seconds.

Anyone just looking casually at that video would have said that the gorilla was the most dramatic and noticeable event in the whole sequence. But Chabris and Simons showed this short film to thousands of people, and about half of those who were counting the team's passes didn't even see the gorilla. They were so closely focused on what they were observing that everything else was edited out of their awareness. In fact, when they had a second viewing of the original video, they were amazed, and some of them even insisted that it wasn't 'their' video – that the gorilla really hadn't been in the footage they had initially studied.

That study shows us just how powerful mental set can be – so much so, that it can make us blind to the obvious even when we're looking straight at it. A replication of the gorilla study using eye-tracking technology revealed that the people who had missed the gorilla and those who had seen it had both spent the same amount of time looking directly at it. The former group just hadn't registered it. The researchers called it **inattentional blindness.**

Inattentional blindness: when someone entirely fails to see something that is in plain sight because they are focused on something else.

Throughout our lives, we are already half-prepared for what we will encounter, so our minds are set, prepared to deal with it. But, as we've seen, our mental sets can lead us into habitual but inappropriate ways of tackling problems. And sometimes, they can even cause us to not notice something that's right in front of our eyes.

Didn't you see that bike?

Inattentional blindness is far more common than we might imagine. For example, it's the reason for most motorcycle accidents. The motorists involved in these incidents almost always say that they 'simply didn't see' the motorbike coming up behind them. This is because they are focused on other aspects of road use – for example, what other cars might be doing – and are just not expecting to encounter a motorbike. It's also noticeable that drivers who have had the experience of riding a motorcycle on the road don't tend to have this type of accident – they are far more prepared to see a motorbike than someone who has never ridden one.

3
I SAW THIS ADVERT ...

Have you ever added up the number of choices you have to make in a single day? I doubt it. Modern life is so full of choice – what to wear, what to buy, what to watch, what to eat, how to travel – that the list is endless. How on earth do we manage it?

If we really considered all the options available to us, we'd never have the time – or brain space – to think about anything else. So when we have to make a choice, we use shortcuts, or heuristics, to streamline our thinking and help us to make decisions quickly.

Sometimes, we just go for the option that comes most easily to mind. The **availability heuristic** is a mental shortcut that does just that. If you want to order a takeaway, for example, you're likely to phone the delivery company whose advert comes most swiftly to mind – either because you've seen it most recently or because it was so particularly striking that you haven't forgotten it. It's readily available in your thinking and choosing that one means you don't have to consider lots of alternatives. Like other heuristics, the availability heuristic saves you from having to do much mental work.

Availability heuristic: the way that we tend to choose whichever option comes most readily to mind.

Mental availability doesn't just relate to adverts. It may be the result of personal experience – for example, deciding only to look at a certain brand of computer or personal phone because that's what you've used in the past. Or it might arise from an awareness of consequences. For example, if you found the perfect present for a friend in a particular market, you might favour that market as the place to begin looking for another gift. Or if a family member had a disaster with a specific make of washing machine, you'd rule out that brand from your options when you were choosing your own.

Availability sort of works on the principle that if you can think of it, it must be important. Which means it can also distort our thinking quite a lot. If you ask people for their views on what the most common crime is, for example, the majority will say murder. We almost always overestimate how frequently it occurs because murder is the crime we hear most about, through both the news media and TV dramas. It is actually quite rare in modern societies, but it gets a lot of publicity, so it is more cognitively available than other offences – it comes most readily to mind. We have to think much harder to identify other types of criminal activity – unless they are part of our recent experience, of course. Asking someone who's been recently burgled about what they think the most common crime is gets quite a different result!

The availability heuristic can lead us to make silly or inappropriate decisions by not really considering better solutions. But it can also, sometimes, work in our favour. It can, for instance, help us to deal with a tricky or dangerous problem if we can bring to mind how someone else dealt with a similar one. But it's a powerful heuristic in our thinking – and one which advertisers and promoters, of course, are well aware of.

Another heuristic that saves mental work, particularly when we're considering more complex decisions, is **satisficing**. Essentially, satisficing is settling on the first solution that satisfies your minimal requirements. A friend of mine was recently looking for a flat to rent while she was away at college. She had a long list of properties to view, but wasn't keen on the whole process and just wanted to get it sorted and go home.

So, instead of checking out all the options, she plumped for the first flat she looked at – and deeply regretted it later when she realized she could have had something much nicer for the same rent. That's satisficing. It saves mental work – but, as with many heuristics, it doesn't always result in the best possible choice.

Satisficing: selecting the first available option which satisfies the minimal requirements for the task.

We use other heuristics too – psychologists have identified over a hundred of them, but we're not going to cover all of them here. We might plump for a particular set of options because we've come across them most recently, so they seem relevant. Or we might choose the ones that seem most appropriate for people of our own age. There's the status quo bias, which doesn't have anything to do with rock bands, but is all about our general preference for keeping things as they are rather than making choices which will change them. Heuristics can lead us to make mistakes, but they can also sometimes show us the most appropriate way to go. But they are all ways in which we can reduce the amount of cognitive work we do and streamline our thinking.

4
IS THAT NORMAL?

We make judgements about what is typical or what most people would do, but these are frequently based on faulty logic. And, quite often, we are barely aware of the decisions we are actually making. We just think about things and assume that what we are about to do or have done is just what anyone else would do or have done. It's only normal, we believe. So if we're making a decision or thinking about something, we fall back on what we consider to be the most likely, or typical, explanation.

Many of our decisions are based on limited information. At such times, we're often influenced by what we think of as the most typical example. Then we use the **representativeness heuristic** – a mental shortcut where we opt for the item or choice which seems to be most representative, or typical, of the whole set. For example, if we saw a stocky, middle-aged man and a thinner, younger man, and we knew that one of them was a builder and the other a librarian, we'd automatically assume that the stocky one was the builder and the other was the librarian. It would almost be a no-brainer: we'd be surprised to find that we'd got them the wrong way round.

This is an obvious example, but the representativeness heuristic can trip us up in more subtle ways too. The trouble is that essentially it's just about assessing similarity – whether that choice is like other examples we've come across, or fits with our existing stereotypes. It's an easy option that doesn't take a lot of thought. But it can lead us astray – just because something appears similar, it doesn't mean that it really is.

Representativeness heuristic: a common error, which is our tendency to judge things by comparing them with similar ones we have already encountered.

We use many other heuristics – shortcuts – in our thinking. On one occasion I was camping with a companion in the Scottish Highlands. We had made friends with the people in the next tent, who were Highlanders themselves but from another area. A helicopter flew over the campsite and when our new friends commented that it was probably out to rescue someone, we were both surprised. Living in the city, we had simply assumed that it was engaging in police surveillance, which was our everyday experience of helicopters flying overhead. We had each used the **base-rate heuristic** – another shortcut in our thinking – which draws on our own experience to explain what is going on.

The base-rate heuristic works well in most situations, but sometimes it can be a real problem. One late-twentieth-century study, for example, found that West Indian men were more likely to be diagnosed as mentally disturbed than white men, even those who were similar in intelligence, personality and family background. It turned out to be a cultural difference and nothing to do with abnormality. West Indian culture encourages energetic, forceful ways of expression, but the more restrained (or repressed) white psychiatrists found that uncomfortable and not what they were used to seeing – except in highly disturbed people (who formed their base-rate information). So they tended to diagnose West Indian men as disturbed or abnormal too, when of course they were nothing of the kind. What we think of as normal is often just what we ourselves are used to.

Base-rate heuristic: a mental shortcut that causes people to assume that what they already know is representative of, or explains, new things that they encounter.

Another of the heuristics that we use in our thinking is called myside bias, and it's one of the most common biases of all. Understandably, we tend to evaluate ideas or things we come across in terms of our own beliefs or preferences, and this leads us to be biased: we're more likely to choose the option which fits our own ideas or inclinations most closely. But it can lead us to make serious mistakes, because it means that we ignore other options and interpret ambiguous information in the way that we want.

Myside bias turns out to be extremely difficult to overturn. When we have strong preferences or beliefs, our basic tendency is to reject information which will challenge them. We find reasons why it is unreliable or its source is suspect, or sometimes we simply don't notice it – we filter it out. Does that mean we never change our minds? No, but we might take a lot of convincing. And with the way that almost any beliefs, no matter how weird,

A different climate

When the Disney corporation was planning a European outlet to supplement the successful US theme parks Disneyland in California and Disney World in Florida, the main options were narrowed down to a site near Paris in France, and an area close to Madrid in Spain. The decision-makers considered from the outset that a location near Paris would be more easily marketed and would seem more glamorous than the Spanish one. Despite being provided with considerable information about both places, including climate and weather data (it was far wetter in Paris), they chose the French option for their new development. The theme park was designed to mirror the two popular US sites, both of which had high levels of sunshine and very little rain. So, when Disneyland Paris first opened, they found that the visitor experience was marred by people having to endure wet and cold conditions as they queued for the rides, and that the site design had a general lack of shelter or provision for inclement weather. This myside bias in decision-making cost the company millions, as it had to adapt the infrastructure of the French establishment in order to deal with something that could have been acknowledged at the planning stage. The Spanish location might have appeared less glamorous superficially, but it would have been more cost-effective in the long run.

can be reinforced by others on the internet, it can result in our living in a cognitive bubble of our own making, which isn't easily penetrated by contradictory information.

5
IS THAT TOO RISKY?

You'd think, wouldn't you, that sharing information with other people and reaching joint decisions would be better than just making decisions individually. But that's not always the case. Sometimes, making group decisions can be completely counterproductive – or even disastrous.

When we are in groups, we tend to feel that we, personally, aren't as responsible as we would be if we were acting on our own. So the decisions that the group makes can easily become extreme. Occasionally, groups reach more risky decisions – they decide to take actions which are more challenging or unsafe than they should be. Sometimes, though, they make choices that are too cautious. It's known as **group polarization**: a tendency towards extremes. A lot depends on how the discussions in the group develop. If one person is advocating a risky strategy early on, others may begin to think of even more challenging examples, and that leads the discussion towards reaching a riskier decision. But if someone advocates more cautious approaches at an early stage, this too can influence the direction of the discussion, resulting in a more cautious decision than the group members might have made individually.

Group polarization: the tendency for groups to settle on more extreme decisions than the individual members would make on their own.

There are several reasons for this. One is shared responsibility, but it also has to do with the principles of the group as a whole: if they value safety and security highly, they may be more inclined towards safer options; if they see their company or section as dynamic and challenging, they may be more likely to take risky ones. Another reason is the way that the group members may be trying to impress others, by appearing either more daring or more cautious. But whatever the cause, group discussions can end up polarizing decisions, sometimes to an unrealistic extent.

More seriously, sometimes – in fact remarkably often – groups make decisions which are completely disastrous simply because their respective members have become complacent and don't really take account of information that might challenge their assumptions. History is full of examples of this: the abortive US invasion of Cuba at the Bay of Pigs in 1961 as the CIA tried to oust the socialist Cuban leader Fidel Castro; the appalling decision to launch the Challenger space shuttle against technical advice in 1986 because NASA wanted the publicity from sending the first civilian into space; the collapse of Barings Bank in 1995, and so on.

How (lack of) groupthink saved the world

On at least one occasion, the lessons learned from a groupthink disaster probably saved the world. After the Bay of Pigs fiasco, in which an attempt to invade Cuba resulted in an extremely embarrassing defeat for the US military, President John F. Kennedy adopted a deliberate policy of encouraging argument and challenges among his decision-makers. When the Cuban Missile Crisis of 1962 happened – an extreme escalation of the Cold War which created a very real likelihood of nuclear warfare starting at any time – Kennedy was able to appraise the situation more realistically, and took the chance to back down when the opportunity arose. That, together with the decision of the brave Soviet officer who actually disobeyed a direct order and refused to launch his missiles, giving Kennedy that opportunity, saved the world from an all-out nuclear war. But for several weeks the world's future hung in the balance, and if the US decision-making had been as gung-ho as it was for the Bay of Pigs the previous year, it seems unlikely that nuclear war would have been prevented.

We have another clear, modern example of **groupthink** in the pre-2020 complacency of the vast majority of political leaders with respect to the likelihood of a pandemic of some kind. While they paid lip service to the idea, practically all Western leaders failed to prepare for such a pandemic, despite repeated warnings that one was virtually inevitable. There was a sort of illusion of invulnerability: that it might happen, but probably wouldn't or at least not yet. As a result, when the Covid-19 pandemic did actually hit, they were largely unprepared.

Groupthink: the way that long-standing groups become complacent and make ill-judged decisions because they don't think they can make wrong ones.

The mechanisms of groupthink are well known:

- *Rationalization* – when excuses and justifications are used to dismiss solutions that the group finds unpopular.
- *Conformity* – when the group insists that everyone should conform to the majority view, instead of taking different ideas or doubts seriously.
- *Stereotyping* – when opponents are stereotyped or ridiculed, and this is used as an excuse for ignoring their arguments.
- *Illusion of unanimity* – when people hide their true opinions in order to avoid being ignored or ridiculed.
- *Mindguarding* – when some members directly censor or shut out contradictory ideas or opinions.
- *Self-censorship* – when members of the group who disagree keep quiet about it rather than speaking up.
- *Illusion of invulnerability* – when the group or committee

acts as if nothing could possibly go wrong.

- *Illusion of morality* – when the group assumes that all its decisions and actions are good and right.

Groupthink is a really easy – and common – trap for a long-term or powerful group to slip into because it can only be properly challenged by encouraging open argument and contradictions from its members. Many leaders find that uncomfortable. But as history shows, groupthink teaches painful – often terminal – lessons. It can only really be avoided by encouraged debate, challenges and realistic discussions in decision-making. A board or governing body in which everyone appears to agree about everything is at serious risk of groupthink – and its consequences.

6
I CAN'T GIVE UP NOW

Unless you're one of those lucky people who only drive new cars, on at least one occasion you've probably experienced the painful realization that your car is too old and no longer worth repairing any more. As a car ages, more and more things can go wrong and need fixing. At some point, the owner needs to decide: is it worth getting this latest issue looked at or is it time to give up on this car and find another? The problem is that a lot of money has already been spent on the car, and scrapping it makes it seem as if that money has just been wasted, which makes it very difficult to choose the best option. It's a problem known as **entrapment**, when a person gets trapped into making the wrong decision just because they've previously invested so much.

Entrapment: an increasing commitment to an ineffective course of action because of the resources that have already been invested in it.

In the car example, it could be called 'throwing good money after bad'. Although we might feel that what we have spent to date will be wasted if we don't do something to try to make it worthwhile, the reality is that the money has gone: throwing even more of it at the problem simply wastes more cash. Realizing when it is time to give up and get out of the trap is important.

Another name for entrapment is the sunk-cost fallacy. Sunk cost is an economic term for what we have already invested – but a sunk cost isn't necessarily about money. It's not uncommon, for instance, for someone to have spent several years studying for a particular degree only to find themselves attracted by a different career. Those years are the sunk cost and the fallacy enters play if they decide to stick with the original career plan, which no longer seems attractive, simply because otherwise those earlier years would have been 'wasted'. So they find themselves stuck in a job they don't enjoy, wasting even more years until they finally

Getting out of a bad situation

The car example is just a small one, but entrapment can happen in other, more serious ways. A notable case of entrapment was the US war in Vietnam, which lasted from 1955 to 1973 and continued long after it had become apparent that the US could not possibly win it. But the Americans had invested vast resources in the conflict, and it had cost huge numbers of lives on both sides. Ending that war was seen by senior military and politicians as undesirable because it might be interpreted as an admission of weakness internationally, and no politician wanted the US to lose face. In US domestic politics, the main arguments centred around what the conflict had already cost, and how all those lives would have been wasted if the US gave up and pulled out. So the war carried on, costing even more lives, until eventually public opinion began to change and President Richard Nixon concluded that it had to end.

have a midlife crisis and embark on a complete change of career! We might decide to stay in the cinema to continue watching a film we're not enjoying because we've paid for our ticket. Or stay too long in a damaging relationship because we've tried so hard to make it work. There are many different ways that the sunk-cost fallacy can affect how we think and behave.

Entrapment and the sunk-cost fallacy also link with another tendency in decision-making called **loss aversion**, a term used to describe the way we are more sensitive to potential losses than to potential gains. They make more of an impact on our thinking. We see this often in TV game shows, where contestants may have already won a large amount of money but feel that they have failed if they don't win the really big prize. Their sense of loss cancels out the pleasure they should feel at what they have already won. We're particularly vulnerable to loss aversion during times of depression – in such moments, we are generally more likely to focus on negatives than on positives, so it becomes harder to be philosophical about losses and concentrate on what we have achieved.

Loss aversion: the tendency to be affected more by potential losses than potential gains.

Both entrapment and loss aversion can affect our thinking, in particular by leading us into making unwise decisions. If we are alert to them, we can dodge being caught – we can set ourselves limits, for example by deciding how much we are prepared to pay to keep an old car on the road, or by concluding that enjoyment or job satisfaction are more important in life than avoiding 'waste'. But the key thing is to be aware of when it is happening.

7
DOESN'T EVERYONE AGREE?

As human beings, we all have our likes, dislikes and emotions. There are things that we feel good about, things that we absolutely hate and lots in between. That's just a normal part of being ourselves. But our emotions, feelings and opinions have a powerful influence on our thinking.

For example, we all prefer to hear things that will strengthen our existing views. It seems to affirm that our ideas are correct. We also tend to avoid or reject information that will challenge our opinions. That's called the **confirmation bias**. It means that anyone trying to change our views, or trying to get us to accept new and different ideas, needs to be more convincing than someone who is telling us something that fits with what we already think. We're just more ready to accept ideas that tie in with our own viewpoint.

Confirmation bias: a tendency to be more receptive to information that confirms what we already believe or want to believe.

The confirmation bias also means that we generally assess how worthwhile statements or ideas are – judging whether they are true or important – in terms of our own beliefs. In chapter 4 we looked at myside bias, which shows how we don't just prefer information that fits in with our ideas, but that we also think it is more worthwhile or significant. And a third tendency in our thinking is the way that we choose what we want to pay attention to. We don't just reject unwelcome information – we actively avoid that which is contradictory to what we believe or prefer. This selective exposure means that, by choosing what information we are exposed to, we can remain comfortable in our beliefs. We don't have to face up to the cognitive disturbance that might be caused by contradictory facts.

It's not just our beliefs and opinions that affect our thinking in this way. Our moods do too. We've all had the experience of feeling down now and again. Even if you're generally pretty cheerful, there are some days when you don't feel as bright as others – and many people experience times when they really feel quite low. We also have the opposite – days when things seem just fine, and we feel quite OK and sometimes happy. Psychologists have found that the emotional state we are in directly influences how we think and what we can remember. If we're in a low mood, we are more likely to remember unpleasant or depressing events; if we're feeling happy, we tend to recall more positive things. In a relationship, for example, someone who is angry or in a bad mood will focus on memories about their partner's inadequacies or irritating behaviours; but if the same person is in a happy or contented mood, those same memories may be totally forgotten. That's known as **state dependency**. The emotional experience creates a physical and mental state, which forms a context for our memories that directly affects how we think in that moment, and also the decisions that we make.

Conspiracy theories and the web

The big attraction of conspiracy theories is the way that they negate the idea that things sometimes just happen, or that people in responsibility may really act out of simple ignorance. Instead, they confirm people's suspicions that 'they' are responsible for whatever bad thing has happened or is going on. In that sense, they are one of the most common examples of the confirmation bias in modern society. The dramatic rise in the amount and range of conspiracy theories has everything to do with the rise of the internet, and how easy it has become for people with common ideas and interests to link with one another. The internet has had many social benefits, of course, and has done a great deal to reduce loneliness and isolation. But it has also meant that people with extreme, and sometimes dangerous, beliefs have found it easy to make contact with others holding similar views. This reinforces the tendency to confirmation bias by providing support for their beliefs from others, and making it easier to reject disconfirming information. Over time, this means that people can become completely surrounded by media input that confirms their beliefs and isolated from any contradictory information. That leads to even more extreme practices, and has even been shown to encourage extreme social violence from particularly disturbed individuals.

State dependency: the way that our emotional state or mood primes our thinking and memory, affecting our choices and what we can remember at that time.

It isn't perfectly balanced, though, in that we have a slight tendency to focus more on negative or unpleasant memories than on positive or pleasant ones. In part, this comes from our social natures: most of us prefer to get on with people, so we are particularly tuned to react strongly to events which break the social consensus. That's important for any social animal. Most of our interactions with others are either pleasant or neutral, and we just accept those as normal. But just one aggressive or nasty encounter can affect our whole day, wiping out any positive experiences that same day has given us. That's the experience we remember.

You'd think, wouldn't you, that being happy and contented would mean that you were also prepared to splash the cash? That happy people would be more generous than those who are miserable? Well, you'd be partly right: happy people are certainly more prepared to donate to charities or to give money to needy people on the street. But when it comes to paying for a commodity or an event, that's not the case. There is what researchers have called the misery-is-not-miserly effect, which is the way that people who are sad or depressed are generally prepared to pay more for such things than others will. Perhaps it's because they feel they are entitled to spend more to cheer themselves up, maybe it's comfort spending or perhaps they simply can't be bothered to look at the alternatives. But it's something you might like to bear in mind the next time you're in a low mood and heading to the shops.

2

MAKING JUDGEMENTS

This section of the book is about the way we evaluate and assess the world around us. Our world includes other people, and we're often quite concerned about how they see us and what they might be thinking of us. We also justify to ourselves why things have happened, or why people – or even ourselves – acted in particular ways, and those attributions affect both how we understand our world and our motivation. We need, for example, to feel that we have some control over our lives: a constant sense of helplessness can even lead us into depression.

Our attributions also affect our judgements. We have a strong tendency to judge other people, although we don't assess their behaviour in the same way that we review our own. And we make judgements about responsibility, but they too are not always logical.

8

DO FIRST IMPRESSIONS MATTER?

We form impressions of what people are like very quickly – practically as soon as we meet them, in fact. They're not always correct – studies vary but tend to show an accuracy rate of about 60 per cent – and the reason for the meeting helps to shape them too. Your first impressions of someone could be quite different if you were a reporter about to interview them for a magazine, a prospective first date or a potential employer. But regardless of the context, they are always powerful.

So how do first impressions work? Well, they begin with the mental set that we looked at in chapter 2. We all have life experiences and knowledge about other people, and this comes together to generate expectations about what particular individuals will probably be like. How would you visualize a librarian, for instance? And would that mental picture be different from an engineer? Or a farmer? Now ask yourself: have you actually met any librarians, engineers or farmers in real life? And if you have, did they really fit those mental images that came into your head?

Probably not. We all have general stereotypes about people and their occupations, and we also know that in real life people are often quite different. But if you're going for a job interview, for example, that kind of idea will also be in the mind of your interviewers, helping to shape their expectations. Almost all of us have to go through interviews in the course of our lives – and if it's for a new job, mental set can make a huge difference to just about everything.

Your interviewers will be mentally prepared in other ways too. They will have looked through your application and CV, and how you described your previous experience, your interests and your ideas will also have developed their mental set. So, even before you come through the door, they will have formed their own impressions from your application, as well as their thoughts about the ideal candidate for the job.

This gives you a choice: do I want to fit their perceptions of the typical person for the job so closely that they think I am perfect for it? Or should I present myself as so different, challenging and exciting that they really want me as part of their team? Only you can make that decision.

But whichever you decide, the first impression that you make as you walk through the door is going to be all-important. Studies of **primacy effects** have revealed that the first piece of information we receive has much more effect on our judgement than any details we come across later.

Primacy effect: the way that the first few items of information in a list or set are the ones we are most influenced by and are most likely to remember.

In one study, for example, they showed quiz contestants answering thirty questions and getting fifteen of them right. Some got their correct answers towards the beginning of the quiz, while others gave them later on. When a group of observers was asked to review how well the contestants had performed, they consistently overestimated the scores from those whose correct questions came early, thinking they had eighteen or twenty answers right, and underestimated the others as being between ten and twelve.

Creating a good first impression can do a lot of work in your favour – being cheerful and smiling as you enter a room has a very different impact compared to looking serious and determined. Positive information in a job application helps too, especially if you've done something worthwhile or in an outstanding way. That's because it can generate a **halo effect**. High-achieving athletes, for example, often find it relatively easy to find work in other fields when they retire from sport. Their achievement and dedication impress prospective employers and cast a positive gloss over their other abilities. So their interviewers make allowances and may even overlook lack of experience, which would make an ordinary person seem unsuitable for that job. A halo effect provides a major advantage, in the same way that a bad first impression puts those being judged at a serious disadvantage. It's part of how our thinking is shaped by first impressions.

Halo effect: the way we make positive judgements about people from our knowledge of unrelated positive qualities that they have.

Good at everything?

The halo effect has a long history. In 1920, American psychologist Edward Thorndike asked commanding officers to evaluate their soldiers on traits such as physical qualities, leadership, intelligence, personality and good looks. He found that good-looking people were seen as kinder and more generous, while those judged to be more intelligent also received higher personality and leadership scores. This was the halo effect at work. It means that being physically attractive can make you seem more creative, while a pleasant personality implies that you are competent and capable. It's a common error in judgement that we are all likely to make, and it's consistent: recent studies have shown how a high intelligence score is taken to imply friendliness and sense of humour, and how a 'healthier' appearance at a job interview is assumed to imply that people have better leadership qualities.

So it really matters how you come across in those initial few seconds. It can affect how warmly the other person interacts with you (which also influences how you respond to them), the questions they ask and how much they notice information that challenges their impression. Experienced interviewers are aware of this – it's why you sometimes have several interviewers and scripted questions. Always keep in mind that a bad first impression will be hard to overcome, whether it's during a job interview, a first date or simply meeting a friend of a friend.

9
HOW WILL THEY SEE ME?

We know that it's important to make a good first impression. But how should we go about it?

We all have multiple versions of ourselves – we're different with our friends, our work colleagues, our close family and with strangers. That's completely normal – we'd be a bit weird if we were exactly the same all the time. But we also choose the image we want to project to other people. In the modern world, that's normal too – Instagram, YouTube vlogs and other social media means that image projection has become a regular part of everyday living.

The key bit, though, is projecting the right image. Media people are well practised at this and so are professional vloggers. The rest of us might find it a little trickier. That's partly because conveying the 'right' image means knowing what will be seen as 'right' in that context and by those to whom you're projecting. You might feel that being a goth is a totally acceptable life choice and image, and in an interview for an art course that would be fine. But if you're interviewing for a financial services trainee role, dressing as an extreme goth could lose you the job before you even open your mouth if it challenges the interviewer's assumptions about what a suitable person would be likely to wear.

We all have our preferences and a sense of the clothes that make us most comfortable. And what we habitually choose to wear can become an important part of our self-concept – so much so that a different style can feel as though it isn't our 'true self' at all. But if we're dealing with people who we haven't met before, their very first reaction will be to our appearance. So it matters. Someone who identifies as a goth is just as likely to be as competent and professional as anyone else, but if they were going for an interview, and being aware of how people's stereotypes work, they might choose a middle road by wearing black and retaining some other aspects of the look while toning down the rest. The aim would be to project a competent and reassuring image to an interviewer, no matter what that interviewer's stereotypes or beliefs might be.

Mirroring: the way that we unconsciously copy the posture of the other person in a conversation.

Body language is another aspect of image projection – and it's an important part of how we think about people. But it isn't necessarily conscious, and one of our least conscious ways of using body language is **mirroring**. When we're listening closely to someone, or in a personal conversation, we often unconsciously copy that person's posture. It's a sign of empathy, signalling that we are in tune with them and understanding what they are saying. We do it unconsciously and the other person reads it unconsciously, but it has a strong effect on how we interact.

Dressing for respect

A uniform is much more than just a set of work clothes. It directly affects how we deal with other people and also how we see ourselves. In a study of Ghanaian student nurses, their comments about their uniform showed how their professional appearance supported them in conveying a knowledgeable and trustworthy demeanour.

> *'I look very neat and tidy when wearing my uniform.'*
> *'My family love to see me in my uniform.'*
> *'The uniform assures my patients that I know what I am doing.'*
> *'Patients will respect and trust me more while I'm wearing my uniform.'*

Wearing a uniform expresses a range of social meanings, not only to people interacting with the uniform-wearer, but also to the wearers themselves.

Individuals with a high level of social anxiety keep a tight control over their posture, and often don't mirror in the ways that others (unconsciously) expect. So they can project an image which is a bit cold and formal. That makes it harder for them to get along with other people – which doesn't help to reduce their social anxiety either. Consciously trying to use mirroring appropriately has been shown to help reduce social anxiety. On the other hand, mirroring isn't always a good thing. One study, for example, showed that too much mirroring in job interviews means you come across as untrustworthy, weak and/or manipulative. When you think about it, that's not surprising: mirroring is a signal of empathy, and is that really appropriate in a job interview?

Politicians have special training in image projection. They know that how they come across to the public has a huge effect on how people will think about them, and – most importantly – whether they will get votes at election time. It's not just about appearance, either. Female politicians, for instance, are encouraged to adopt lower voice tones, so they sound more decisive: things said in a high-pitched voice are often interpreted as indecisive or weak. Politicians are trained to avoid other aspects of **paralanguage** too – hesitancies such as 'um ...' and 'er ...', and rising pitch at the end of sentences, which implies that the person is unsure and asking a question rather than making a statement.

In the end, it's all about which of our multiple selves we want to be seen by others. Our projected image, if we want to keep it up for any length of time, needs to reflect a real aspect of our own selves (only maybe a little more polished up and without the dodgy bits). And different jobs bring out

Paralanguage: the non-verbal cues that are used in speaking to add meaning to the words being used.

different aspects of personality: the social skills required of a supermarket checkout assistant are not the same as those needed by a lab technician. So which side of you do you want to project? Is it the cheerful socializer, the team player, the keen video-gamer, the quiet thinker or the reassuring helper? We can be any or all of these and a lot of other people besides: it's your choice.

10
I CAN DO THIS!

Are you a trier? Or do you give up easily when things get difficult? Perhaps your answer is simply, it depends. If you said that, chances are that you would only be half right. You'd be correct in saying it depends, but probably wrong in what you think it depends upon. You might imagine that it depends on the situation or on the type of problem that you're puzzling over. But what it really depends on is you.

It's all about our **self-efficacy** beliefs – what we believe we are capable of. We each have our own ideas about what we can and can't do, what we're good at and what we are just hopeless at. We also know if we have the ability to learn new skills, and which ones we might be able to grasp more easily than others. Each element adds up to how effective we see ourselves being in any given context. And that makes all the difference in terms of how we go about tackling challenges.

Self-efficacy: how effective someone believes they are likely to be in working to achieve a specific goal.

Take maths problems, for example. Any maths teacher will tell you that some children simply won't accept failure – if they don't solve a problem at first, they'll keep trying until they do. Others, though, will only have one go at it. If that doesn't work, they just give up. An important part of the teacher's job is trying to get those children to see that it's worthwhile trying alternative approaches. The ones who persevere know that: they believe that if they try hard enough, they'll be able to solve the problem. Those who give up can't see any point in trying any further because they don't think they will be able to work it out anyway. They think they 'can't do' maths.

Those children see the failure as being all about ability: they imagine they simply lack the ability needed to do maths. But interestingly, ability turns out to be much less important than we think. Studies have shown that the children who are most successful at maths at school aren't necessarily the ones who

Mindset

Psychologist Carol Dweck worked closely with fellow psychologist Albert Bandura as he developed self-efficacy theory, and went on to explore how these ideas apply in different areas of life. She found that it isn't intelligence, education or talent which leads people to succeed in their life goals: it's about how they go about facing life's challenges. Dweck coined the term 'mindset' and argued that this is what makes the difference. Some people have a fixed mindset, believing that their intelligence or abilities are simply a hand they have been dealt by nature and not something they can change. This means that they tend to avoid challenges, fearing that they might be inadequate in dealing with them, give up easily when they meet problems and ignore potentially useful negative feedback. Sometimes, too, this can lead them to feel threatened by others they regard as more gifted. Those with a growth mindset, on the other hand, see intelligence or ability as something that can be developed. So they are more likely to embrace challenges and persevere in the face of setbacks, learn from criticism and be inspired by others. More importantly, Dweck and others have also found that people can change their mindset through appropriate training and experience.

start out with the greatest aptitude. The ones who really do well are those with high self-efficacy beliefs – who believe that even if they can't do something right now, they will be able to learn how to do it. So they try harder and achieve the skills that they need. Sometimes, children with a high natural ability become demoralized when they encounter problems that they can't solve easily and they give up. It's those high or **low self-efficacy beliefs** that make all the difference.

Low self-efficacy beliefs: having a tendency to attribute personal achievement to fixed ability or intelligence rather than to effort.

The same applies to real-life problems as well. Many people who begin DIY projects in their homes find that they need new or unexpected skills to complete them. Some choose to give up at that point. They might look for a professional to finish the job or they might – to the exasperation of friends and relatives – just live with the half-finished work. Those with higher self-efficacy beliefs are more likely to buckle down and learn the new skills. They might ask other people they know for help, or find internet tutorials or even classes. But the important thing is that they overcome the obstacle in their way and complete what they need to do.

You may not feel that you can do something, but if you have reasonably high self-efficacy beliefs, you think that you could learn how to do it if necessary. If you have low self-efficacy beliefs, you can't envisage being capable and so you just give up. It's not the same as being self-confident – you might have high self-efficacy beliefs regarding practical skills and low ones concerning paperwork and dealing with officialdom, for example. But they are all about realizing that your abilities aren't fixed and that you could learn or achieve things if you went about it in the right way.

11

WHY ARE YOU SO STUPID?

In an episode of *Doctor Who*, when the Doctor's TARDIS was forced into a human body by an evil alien, one of the first observations the TARDIS personality made was: 'Are all people like this? So much bigger on the inside?' The irony, of course, being that 'it's so much bigger on the inside' is the inevitable remark made by visitors on entering the TARDIS in its regular form. But her comment was absolutely right – we are, all of us, so much bigger on the inside. How we appear to others – even those who know us quite well – covers only a fraction of what we are like and doesn't account for all those thoughts, ideas, knowledge and opinions that we have on the 'inside'.

We all know this about ourselves. What we don't always do, though, is take into account how it applies to everyone else as well. Too often, we make assumptions about people based on our limited external view of them, without appreciating that there may be much more behind their actions than is apparent on the surface. In other words, we judge others quite differently from the way we judge ourselves.

It's simply, to quote playwright Alan Bennett, a question of **attribution**. Attribution is all about the reasons we give for why things happen – for example, whether we see things as being caused by our own actions or deliberate choice, or resulting from circumstances over which we have no control. The process of attribution is really at the heart of how we understand what is going on with other people, and sometimes with the world at large. It has several facets, some of which we'll look at in the next two chapters. But what we're interested in here is about whether we think something has been caused by internal factors or external ones.

Attribution: the process of giving reasons or explanations for why something is the way it is.

Internal factors are things like personality, ideas, opinions or characteristics. We might, for example, attribute a difficulty in sleeping to being worried about something or anxious about an upcoming event. That would be an internal attribution because it comes from something within ourselves. Alternatively, we might ascribe sleep problems to an overly bright full moon or to noise from neighbours or to the temperature being too warm. Those are external attributions – the cause doesn't come from ourselves, but from the outside world.

Them and us?

The fundamental attribution error has been shown to apply at a group level as well as when we are looking at how individuals make judgements. In one study, for example, Taiwanese people were asked to explain the failure of an election candidate. Those who identified that person as a member of their in-group tended to use situational factors, while those seeing them as a member of their out-group used dispositional ones. It has also been shown that people from more 'collective' cultures, such as China, are less likely to make situational judgements about others than those from more individualist cultures, like the US. Researchers investigating this more deeply have found that it's all about the type of situational influence that is concerned. People from collectivist cultures are more likely to make situational judgements if there are social influences or pressures involved, and less likely when they are evaluating purely individual actions.

Most of the time, we attribute our own actions to external factors or situations and the demands they make on us. How we act at work is different from how we behave at home, which again may vary from how we conduct ourselves at a sports event or music festival. That's pretty obvious: those situations are making different demands, so we act accordingly. We make external attributions for smaller things too. If we fall over suddenly, we look for something that tripped us up – an uneven pavement or a tree root. If we miss a deadline, we attribute it to having too many other things demanding our attention. In other words, when we do something that needs explaining, we look for a reason and we usually find it in the situation we're in.

But we don't use the same standards when we're judging what other people do. One of the classic errors in our thinking is known as the **fundamental attribution error**, and it's called that because we do it almost all the time. We regularly attribute what other people do as arising from internal causes. If I scrape my car, it's because there were other things in the way and there was no room to manoeuvre; if my friend Rick scrapes his car, it's because he's clumsy. If I miss a deadline, it's because there were too many other things that needed doing; if my workmate Janet misses one, it's either because she's lazy or she just can't manage her time. Sometimes, the fundamental attribution error can be positive – if I get up early to bake for a charity event, it's because they need the contribution; if Janet does it, it's because she's a really good baker. But whatever the outcome is, we tend to attribute what other people do to internal causes, but what we do to external ones. And most of the time it doesn't reflect well on the other person.

Fundamental attribution error: the tendency to ascribe situational causes to our own behaviour, but dispositional causes to that of other people.

Unfairly, we judge people as stupid, clumsy or incompetent without even looking for alternative explanations as to why they may have acted as they did. If we used the same standards for them as we do for ourselves, and actually looked for external factors to which they might be responding, our everyday judgements would be so much fairer.

12
WHO'S IN CHARGE HERE?

The idea of internal or external doesn't just apply to our attributions. It's also really important in how we relate to the world around us. Human beings have evolved to be able to manipulate their environments – having an opposable thumb makes it easy for us to use our hands to make changes within our surroundings, and having a large brain gives us a huge capacity for learning new things. That's part of the evolutionary heritage we all share. But it also means that control – the ability to direct changes and make things happen – is absolutely fundamental to everyone. And feeling in control really matters. Studies have shown, for instance, that people can put up with an irritating noise while they are working if they feel they could stop the noise if they really wanted to. But if they can't control it, the noise is more stressful and more likely to interfere with their work.

We're not all the same, and we don't all have the same

opportunities either. Some people's lives contain far more stressful factors than those of others. But individuals differ in how they deal with situations, and whether they feel able to influence or control what happens to them. Some have what's known as an external **locus of control** – they see the control of their own lives as something that is quite outside of themselves. In short, they feel helpless – and we'll look at some of the consequences of that in the next chapter.

Locus of control: a belief about whether what happens to someone is something they can influence (internal), or whether it is caused and controlled by external factors.

Others, though, have an internal locus of control. They are the 'if life gives you lemons, make lemonade' people – the ones who can always make the best of even bad situations. They know that they can't influence everything that happens to them, but they can work with the situation they are in and find ways to make it less bad. They believe that, in general, they are in control of their lives and that even if things aren't perfect, their efforts will make a difference.

So whether you have an internal or an external locus of control, it makes all the difference to how you deal with adversity. As we saw with self-efficacy beliefs in chapter 10, believing that you can do something means you try harder to achieve it, and that in turn makes you more likely to succeed. Locus of control

Nice or nasty?

If we see someone as being in control of their actions, it can strongly influence how we behave towards them. In one study, psychologists looked at the attributions made by carers about the origins of challenging behaviour that they witnessed at a day centre. They discovered that the more outwardly directed the challenging behaviour was (for example, if it involved physical or verbal aggression towards staff), the more the carers made attributions of control and hostility, and that in turn caused them to be less inclined to help. Where they viewed a person's behaviour as uncontrollable and not particularly negative emotionally, the carers were far more ready to offer assistance to them. The nature of the attributions that they made about the behaviour directly affected how the carers responded.

is similar to self-efficacy, but it's the whole approach to life in general rather than being about specific skills or abilities.

It's also linked with whether we make internal or external attributions, although it's not the same. Internal attributions are often controllable, but not always. Thinking 'I failed my exam because I just can't do maths' is an internal attribution, but not a controllable one. Thinking 'I failed my exam because I didn't work hard enough' might seem the same, but is what we call a **controllable attribution** – it's viewing the circumstances as being something that I could have done something about, whereas being unable to do maths is perceiving it as originating in a character trait, and therefore (probably) something I can't do much about. So, although internal attributions often mean we see things as controllable, that's not always true.

Controllable attribution: seeing the cause of an experience or event as something you can influence or do something about.

Controllability matters because we deal with practically everything much better if we think we are in control, at least to some extent. And the sense of being out of control is a major source of stress. We are much more relaxed and confident when we feel in control of things. Psychological studies have shown how even having a fake control over a stressor – one that doesn't actually work but the person believes it might – can reduce people's physiological stress levels. It's important to feel in control, even if that control isn't really real.

Keeping – or acquiring – an internal locus of control is hard, though, when life has really beaten you down. Taking control of their lives again is the main challenge facing those who are recovering from addiction or working their way out of homelessness or dealing with other major life traumas. So, while giving a helping hand makes a big difference, and sometimes it's what people desperately need in the short term, in the long term lasting change only comes when people strive to take back control of their own lives. That doesn't invalidate the helping hand, though – when you read the biographies of individuals who have pulled themselves out of these difficult situations, it was often a random act of kindness or assistance that gave them the spur to act for themselves, or kept them going until they were ready to move forward. But it was taking control that really made the change work.

13

HELPLESS AND HOPELESS?

Back in the bad old days (mostly before the Second World War, but also in the 1960s and 1970s), psychologists used to do lots of experiments with animals. Many of them were pretty pointless, but a few did produce some real insights, and identifying what we now call **learned helplessness** was one. It began with the idea that dogs would learn to perform certain actions in order to avoid electric shocks. No surprises there. When the experiment was adjusted, so that no matter what those dogs did the electric shocks were still applied, they found that the animals would simply give up. They became passive and just endured the shock. Again, not surprising. But, more importantly, when the test conditions changed back to how they were originally, so the dogs could again avoid the shocks by performing a simple action, the animals remained passive, didn't try to adapt their behaviour and didn't learn that they could do something to stop the shocks. They were displaying the attributes of what became known as learned helplessness.

Learned helplessness: an inability to take action even when action would help the individual to escape from negative circumstances.

It wasn't long before psychologists noticed the striking similarities between learned helplessness and human depression. When people become severely depressed as a result of continuous and serious life problems – known as reactive depression – they tend to become passive and give up on trying to help themselves. Often they simply don't see that there are things they could do to improve their circumstances. While some people weather their problems, treating difficulties as challenges and immediately looking for ways to deal with them, those suffering from reactive depression don't do that – they don't think there's any point because they believe they wouldn't be able to do anything about their situation anyway.

As they looked further into this idea, psychologists found that seriously depressed people show characteristic patterns in the attributions that they make. They tend to make attributions that are external, uncontrollable, global and stable. As we discovered in chapter 11, internal or external attributions are dependent on whether we see causes as coming from within the person or from the situation, and in chapter 12 we learned that the controllable or uncontrollable dimension is about whether a cause is seen as something the person can do something about, or not. Specific or global attributions refer to whether the cause is only affecting that particular situation, or if it is likely to apply more widely and affect everything else as well; and unstable or stable ones relate to whether the cause of the problem is seen as being just temporary or as likely to persist in the future.

For people suffering from reactive depression, then, their understanding is that things happen because of outside factors, there isn't anything they can do about them, those factors are likely to apply across the board rather than just to that particular situation and nothing is ever going to change. This pattern of attributions is known as a **depressive attributional style**, and it results in learned helplessness – a person disposed to this way of thinking is unable to do anything to help themselves. They can't see any point in trying because they don't believe that anything they could do would work anyway.

Depressive attributional style: a pattern of thinking commonly found in people suffering from depression, in which negative experiences are seen as stable, global and uncontrollable.

Learned helpless isn't permanent. It can be challenged and changed, and recognizing the depressive attributional style has been an important insight for psychotherapists and counsellors.

An uncontrollable infant?

Analysing attributions has been particularly useful in family therapy. In one study, for example, psychologists reported the use of attributional analysis with young mothers who were experiencing emotional problems with their babies and were suspected of aggressive behaviour towards them. They compared how these women talked about their babies with recordings of similar conversations involving mothers whose children offered similar levels of stress – for example, those who had children with severe disabilities or illnesses. What they found were significant differences in the attributions that the two groups made. The mothers suspected of aggression saw their babies' behaviour as essentially uncontrollable, and their frustration arose from their feelings that there was nothing they could do about it. This was helpful to the family therapists, as it showed them how they could teach these young women to influence and control their children's behaviour. As these parents came to see that there were actually things they could do to help, they were able to relax and enjoy their babies instead of feeling frustrated by them.

The psychologist who first identified learned helplessness, Martin Seligman, also showed that it has a counterpart, learned optimism: that we can learn to become more optimistic by adopting a more positive attributional style.

This idea has become a basic approach in psychological therapy. Cognitive therapy, for example, is all about helping people to learn how to think more positively – for example, encouraging them to shift from making uncontrollable, stable attributions to seeing things as being more controllable and more open to change. Cognitive behaviour therapists help people to learn how to change their behaviour as well as their attributional patterns, so that they can develop lasting habits which will benefit them. And family therapists use attributional analysis to identify the sources of damaging or self-destructive patterns of interactions within families. We hear a lot about positive thinking, and positive attributions are at the heart of it.

14
WHOSE FAULT WAS THAT?

How do we assess whether someone is to blame for an accident? We're not always logical when we attribute guilt or responsibility – so much so, that our judgements are often simply unfair. There's a wealth of evidence, for example, about how physical appearance can affect our judgements. Studies have shown how unattractive children are much more likely to be blamed for mishaps than attractive ones. Children realize this – most schools have at least one charming, attractive kid who seems to get away with just about everything. The child is usually well aware of it too, and able to exploit that charm when dealing with adults.

There have also been studies which have shown how juries respond more positively to attractive people than to unattractive ones. It's unfair, but it's very common. So much so that almost everyone attending a legal court, no matter what the reason, takes pains to make sure they look as neat and smart as they can.

When we are making judgements, we draw on a highly developed set of social knowledge: our personal experience, the attributions we make about why other people act as they do, the everyday social scripts which outline how we should act in particular situations, and our schemas, or how we organize our individual knowledge and experience. As we saw in chapter 11, we're much more likely to blame others for accidents than we are to blame ourselves. The fundamental attribution error is to attribute our own behaviour to **situational causes**, but other people's behaviour to dispositional ones. But it's not just about what we think that person is like. It's also how we assess their intentions – how deliberate we think their actions are.

Situational causes: circumstances or constraints which mean that a particular act or behaviour is unavoidable.

Suppose that a vase has been broken, and Sally was the only one at home at the time. There is some question as to whether Sally was responsible for the breakage. We'd take into account her intentions, whether she was trying to help by doing the dusting, whether she had put the vase right on the edge of a surface so that an accident would be easily foreseeable, and so on. We might even wonder whether she had broken the vase spitefully to get back at her sister, who was fond of it. And we'd be more likely to judge her responsibility according to her intentions – whether breaking the vase was a deliberate act – than according to what actually happened. If it wasn't deliberate, we'd just see it as an accident and not really judge her as responsible.

That's a trivial example. But whether we think someone is responsible or guilty depends on other factors too, and one of the main ones is how it all turns out. In Sally's case – well, the vase was broken, that's all. But what about an accident where someone ends up being maimed or even killed? We make very different judgements then. This is called **outcome bias**.

Outcome bias: differences in judgement about the quality of a decision arising from knowing the outcome of that decision.

Most traffic accidents, for example, are foreseeable – they come from people driving too fast or being tired or not paying enough attention to what they are doing. If an accident just results in a scrape to the car or being temporarily stuck in a ditch, we might make an excuse for the driver – they'd been on the road a long time, they were distracted by an unexpected noise or something like that. We certainly wouldn't see their level of guilt as particularly high. But if someone was hospitalized or killed as a result, we would see the driver as not just responsible for the accident, but as actually guilty of a crime – we view the accident as having been easy to predict, and therefore we regard the driver as entirely at fault. The outcome of an accident has a direct effect on whether we judge the driver to be guilty or not.

Bad medicine

Outcome bias is a constant problem for professional practice. We may be aware beforehand that a medical operation carries a slight risk, but if an operation goes wrong we see the surgeon concerned as entirely responsible for the outcome – even if it wasn't really anything to do with how they performed. We accept, intellectually, that professionals are not infallible, but we still assign blame and responsibility to them as if they were – or at least, as if they ought to be – even if it couldn't have been predicted. This is why doctors and other professionals have to go to such lengths to inform their clients about risks, and often have to take complicated precautions which seem unnecessary to an observer.

MENTAL STATES

What we know as thinking is massively complex. It isn't just working out solutions to problems. It includes having ideas, which sometimes come suddenly and after a long period of mental work; it involves planning activities and using learned skills; and it even includes those moments when you might not feel as though you are actively thinking at all, just daydreaming.

This is because quite a lot of thinking isn't conscious at all. Our brains can be unconsciously incubating problems or consolidating skills without our being aware of what's going on. Highly creative people often don't think about what they are doing: they simply do what their 'muse' drives them to do, and their high level of expertise allows them to achieve it. Logic and reasoning are part of thinking too, as are our plans and intentions. How far we achieve those, though, is strongly influenced by how realistic they are, by our habits and even by our emotions.

15

SORRY, I WAS MILES AWAY ...

When we think about thinking, we usually imagine being problem-focused and alert. But a lot of the time, our thinking isn't really like that. We don't spend our whole waking time focused and alert. We are sometimes daydreaming, immersed in a film or a book, or simply letting events drift past us. Occasionally, it can feel as though we're barely thinking at all – just functioning on autopilot. But even during these quiet times, there's still an awful lot going on – we're still thinking, just not quite in the same way.

Take daydreaming, for instance. When we're children, we are given to believe that daydreaming is something we shouldn't be doing. And that's partly true, in the sense that there are certain moments when it's definitely inappropriate. But actually, spending a little part of your time daydreaming has been shown to be good for your mental health. It allows the brain to relax, taking the pressure off the feeling that we should always be alert and taking notice of everything. That state of mind is important for mental health, as a stress reliever, and it also helps us to be creative and solve problems.

When we are daydreaming, meditating or in a similar calm state, it puts our brains into a different mode of operation. You can see these changes clearly on an **EEG** reading – a recording of your general brain activity. Normal thinking, when we are aware of what's going on but not really focusing particularly hard on any one thing, shows very small variations in the overall level of activity of the brain. But those variations become larger when we are daydreaming, meditating or engaged in some other form of mental relaxation. It's as if the brain can roam a bit more freely. Those larger variations are known as alpha waves.

EEG: short for electroencephalogram, this is a general record of electrical brain activity obtained by placing recording electrodes across the scalp.

Interestingly, we can learn to control our alpha waves. You may come across a demonstration in a science centre where someone puts a frame over their head and then attempts to move an object on a screen by brain power alone – perhaps to change the rotation of a globe or to move a pointer. The secret here is not to try hard but to relax. The machine is picking up alpha rhythms as your brain generates them: if you're too excited or wired up, you won't be able to make it work. If you can relax and control your thoughts, though, your demonstration of 'mind power' can be quite impressive.

There's a different pattern created when we are really concentrating on something. Then, our brains show a series of very tight variations which organize themselves into larger groups, or waves, known as theta rhythms. We all have to concentrate from time to time, but continuous and extreme concentration can sometimes produce a sense of absorption, known as **flow**, which is a distinctive and sometimes almost addictive state of mind. The EEG pattern is the same as when we are concentrating hard, but the state of flow is easier and more pleasant, as if the brain is settling into the concentration mode and doing it without effort.

Flow: a pleasurable state of heightened focus and immersion in activities such as sport, art or work.

These three diverse styles of thinking (normal, daydreaming and concentration) – or states of mental activity – are very general, but it's interesting that they involve completely different patterns of brain activity. We're only just beginning to be able to identify which specific parts of the brain are involved in particular types of thinking – we know something about where decision-making and planning are located, for example, and we could probably predict some of the neural pathways that would be involved in deciding to make a cup of tea. But thinking as a whole involves our own unique personalities as well as personal, economic and social experiences. So we'd struggle to predict the brain cells involved in, say, deciding to move house. But we can certainly tell, electronically, whether people are relaxed, concentrating or just generally aware.

Are you in the zone?

Some people find concentrating hard work, but for others it can be a deeply satisfying state of mind. In the case of the latter, they can achieve a state of flow, which is complete absorption in the activity they are doing at the time. They become so involved in what they are concentrating on that nothing else seems to matter. Athletes often describe this state as being 'in the zone': at such times, they feel utterly focused and able to achieve their peak performance. Writers, academics and others attain it through more cognitive activities, but the originator of the theory, Mihaly Csikszentmihalyi, argued that it is something that almost everyone experiences at some time or another. He also argued that these are the moments when people are happiest – there is a feeling of engagement, skill or efficacy and fulfilment which is wholly pleasurable and rewarding.

16
EUREKA, I'VE GOT IT!

Sometimes inspiration comes to us seemingly out of nowhere. Our thoughts may be focused on a particular topic, and then a completely different idea flashes into our mind. Maybe it's a subject that we were thinking about earlier but had forgotten, or perhaps it's a solution or a new approach to a long-standing problem. That's because thinking and awareness don't always go hand in hand. Our brains are hugely complex and there's much more going on beneath the surface than we realize.

This sudden inspiration is sometimes called the **eureka effect**, after the legend about the Greek philosopher Archimedes, who was sitting relaxing in his bath when he suddenly hit upon the solution to a tricky problem he had been asked to solve. The story goes that he jumped out of his bath shouting 'Eureka!' or 'I have found it!' This type of episode, also known as the Aha! experience, happens quite often. You might identify the solution to a crossword puzzle clue, see the answer to a difficult problem or come up with a new idea for marketing your work quite suddenly and all at once.

Eureka effect: a common human experience of suddenly realizing the solution to a problem.

There are four distinguishing characteristics of the eureka effect. The first is the way that it appears quite suddenly, as if out of the blue. The second is how it gives us a smooth, easy solution to a problem. The third is that we feel good about it – it provides a moment of satisfaction. And the fourth is the fact that it's accompanied by a sense of certainty: when we have an Aha! experience, we are sure that the solution we have found is the right one.

Sometimes, insights can come to us in obscure ways. The nineteenth-century chemist August Kekulé always said that he was only able to resolve the puzzling structure of benzene when he had a dream about a snake eating its own tail. When he woke up, he realized that all that confusing data would make sense if the benzene molecules were organized as a ring. That's a particularly dramatic example, but many of us have found that tackling a problem often becomes easier if we 'sleep on it' – in other words, if we stop thinking about it consciously and allow our unconscious minds to work on it instead. Things often really do look better in the morning! We'll come back to this idea in chapter 46, when we focus on dreaming a bit more closely.

Insight: an understanding of the essential elements of a problem or situation.

Insight or sudden intuition doesn't involve conscious thinking. Instead, it comes to us as a result of the mind tapping into long-concealed memories and unconscious knowledge.

And the more knowledge and skills we have buried there, the more likely we are to have this type of insight. Veteran crossword solvers, for example, have far more eureka moments than people who aren't used to doing crosswords – they simply have more past experience to draw on. And, as we've seen, eureka moments are pleasurable, which makes them a large part of the reason why people choose to do crosswords in the first place.

Our cousins the apes

It's not just humans who can experience insight. Back in 1917 the Gestalt psychologist Wolfgang Köhler was conducting a number of different learning experiments with a small group of chimpanzees. On one occasion, he hung a banana high up on the ceiling of a cage – too high for a chimp named Sultan to reach. Sultan made a couple of futile jumps and then retired to a corner, as if sulking. Suddenly, he leapt up and began collecting the boxes that were strewn around his cage, piling them up so that he could stand on them and reach the banana. Köhler described this as a clear example of the eureka effect – the solution had suddenly come to Sultan and he had immediately acted upon it.

17
IS THAT MY MUSE?

All of us experience insights at some time or another. But some people seem to have the capacity to tap into realms of insight or awareness that the rest of us simply don't have. They stand out as particularly remarkable artists, musicians, poets, writers and other really exceptional people. They have an ability to create material which is unusual, insightful and even breathtaking. By harnessing their imaginations, they are able to produce things that the rest of us wouldn't have dreamed of. In short, these are the people we think of as being especially creative.

Creativity often seems like a magical, unknowable quality. Sometimes it feels like that to the artists concerned as well. But studies of the creative process tell us another story. It doesn't in any way detract from how special those people are, but it does help us to understand how they are able to be so extraordinary.

Obviously, it's difficult to study the creative process at work. There have been many attempts to measure creativity – to develop ways of identifying people who can think that little bit differently. These efforts have been relatively helpful in identifying those who can, for instance, find novel solutions to problems, or even training other people to do that better.

The efforts to measure creativity, though, don't really come close to what we would interpret as the real thing, as shown by high-powered artists, musical composers or scientists. But interviews and biographical studies of truly creative people do

What's the difference?

Highly creative people show a level of expertise in their fields which is similar to that shown by other types of experts, whether they be musicians, mathematicians, scientists, or even chess players. There have been several studies of what makes the difference between an expert and someone who is just learning – a novice – and these five differences have been identified:

- Experts are better at remembering things that are relevant to their area.
- Experts go about solving problems differently. They are able to draw on approaches or techniques which a novice simply doesn't know.
- Experts understand problems in a different way to novices. Novices tend to focus on the elements of a problem, while experts are more likely to see it as a whole.
- Experts know more than novices do.
- Experts have had much more experience, which has allowed them to practise their knowledge and apply it more extensively.

tend to show a pattern in the creative process. The first part of this pattern is skill acquisition. What all of these individuals have in common is that they have spent many years developing their techniques, to the point where mastering drawing a sketch, painting a picture, writing, doing maths or playing a musical instrument comes as second nature to them. Practice, as all musicians know, is the key to acquiring real expertise; and the same applies in other creative pursuits as well.

So that's the background. But when it comes to producing a work of art, or other examples of creative genius, the studies show that, apart from acquiring the skills, creativity involves two more distinct stages. There is a period of unconscious **incubation**, when nothing much seems to be happening and the person concerned is not really doing any conscious thinking about it. That ends with a sudden inspiration, producing a bout of activity as the creative individual paints the painting, composes the musical score, writes the poem or produces whatever it is they were unconsciously working on. This sudden inspiration is a form of insight, but it's a rather special one. Those concerned often describe themselves as feeling 'driven' to articulate their vision.

Incubation: a period of unconscious cognition, during which aspects of a project are worked out, but involving mental activity of which the person is entirely unaware.

The stages of incubation and inspiration are distinctive in creativity, and we can often assume that they are the elements that matter. But the evidence shows that creativity is also firmly based on a high level of **expertise** in the creative medium. Creative artists are also technical experts: they are so familiar with the tools of their trade that they don't even have to think about producing the effect that they want to convey. They can have an idea and realize it immediately. It is those levels of expertise that distinguish them from novices or those of lesser talent.

Expertise: a high level of ability that is knowledgeable, extremely competent and doesn't require conscious thought or effort.

Most of us – even exceptionally talented people – could never achieve the type of creativity shown by genius artists like Renoir, musicians such as Beethoven, writers like Shakespeare or scientists such as Einstein. Nor would we expect to. But there are many levels of creativity. While society includes those who we respect as being highly creative in their respective professions, there are also many people reaching remarkable heights of creativity in their hobbies – whether that be railway modelling, cake decorating, felting or even knitting. For those people, too, it's the development of expertise in the basic skills that has freed up their imaginations, allowing them to realize their visions.

18

HOW DID YOU WORK THAT OUT?

When we think about thinking, we tend to focus on how we work out solutions to problems. But of course, there's much more to thinking than that, and even our problem-solving can be surprising. We tend to see this kind of thinking as essentially rational and logical, but the way human beings go about solving problems often isn't very logical at all.

Logic is useful when the problems we are dealing with are very clearly bounded – like calculating percentages or doing other types of mathematical calculations. But when it comes to problems that are about human behaviour or even the natural world, logic isn't always that helpful. A useful example is how we would react to someone telling us: 'If it's raining on Sunday, I'm going shopping.' If we met that person in a shopping mall that Sunday, we'd be likely to conclude that it was raining. To us, that sounds logical, but it isn't really – or at least, not according to **formal logic**. That person didn't say anything at all about what they would do if it wasn't raining, so as far as formal logic is concerned, the weather could equally well be fine.

Formal logic: the abstract study of propositions or statements in terms of their strict meanings and what can be mathematically deduced from them.

The difference between our assumption and formal logic is that we take into account the human meaning of what was said, rather than simply processing it as a formal logical statement. As human beings, we draw on a number of cognitive shortcuts, some of which are so deeply part of how our minds work that they are even evident in babies. Take the way that we tend to assume causality. If something is immediately followed by something else, we tend to assume that the first has caused the second. Even infants have shown this tendency – for example, if an infant sees an object moving and then passing behind a screen, it tends to look at the place where the object would normally emerge. If it comes out somewhere else, the baby spends a lot longer looking at the object, as if studying it to make sense of what has happened. As adults, if someone takes a phone call and then immediately leaves the room, we assume that they are responding to the call – even though the two actions might be entirely unconnected.

That might be plausible, but we apply the same assumptions even to videos of simple shapes moving about a screen: if a square moves towards a circle and then the circle moves away from the square, we assume the square has chased it away. If we were computers, we wouldn't make that assumption. But as human beings, we are entirely justified in doing so – in real life, that's what we would expect to be happening.

We're also much more likely to choose options that are familiar to us. There has been masses of research showing how familiarity affects our choices – even to the point where those choices would seem obviously wrong to someone who wasn't involved. It's just easier for us to think about things that we already know.

Despite – or perhaps because of – these logical errors, our

But is it really brainstorming?

Brainstorming is a method that has been used to help people to think laterally. It's become a popular buzzword to describe any type of ideas-generating session, though many are not really brainstorming at all. True brainstorming has two very distinct stages. The first is an uncritical ideas-generating stage – and it's the word 'uncritical' which really matters here. Everyone in the group is encouraged to suggest anything at all that occurs to them, no matter how absurd the ideas that have been put forward may seem. All those ideas are collected and nothing is rejected. It's not until the second stage that every suggestion is evaluated and assessed according to whether it is practical. The reason for this is that people often feel a little self-conscious about proposing things they think are a bit silly and may not say them for fear of instant rejection. But ideas which appear unusual or even impractical at first can sometimes turn out to be really good, so separating the two stages of brainstorming is important in making sure that everything gets considered and nothing is rejected too quickly.

thinking tends to be fairly single-minded. We generally focus on one solution to a problem and keep on it unless it's completely invalidated. But if you are faced with a new situation where standard solutions don't work, or if you are in an advertising agency, for instance, and looking for entirely new approaches to promoting your product, you and your team might be in need of ideas that are really different – which are 'outside the box'. You have to encourage **lateral thinking** – thinking which doesn't go in straight, logical lines, but rather explores different possibilities or pathways that lead to some surprising or extraordinary solutions.

Lateral thinking: taking an entirely new and different approach to a problem.

Lateral thinking was first identified as such by Edward de Bono in the 1960s. He described it as the best way of responding to the challenges of a rapidly changing modern world. Recovery after the Second World War was largely achieved, Western economies were booming and a new generation of consumers was becoming influential – as a result, conventional ways of thinking were being challenged in all sorts of walks of life, and there was a general trend towards novelty. The idea of lateral thinking mirrored that social mood, by throwing off conventional constraints and adopting entirely new perspectives. As the consumer society continued to develop, the need for novelty also continued and lateral thinking, like brainstorming, has become an everyday part of commercial decision-making and language.

19
DO YOU KEEP YOUR NEW YEAR RESOLUTIONS?

Thinking isn't just about the present and the past – even though they may occupy a lot of our attention. We also think about the future. We have intentions to do things and we make plans for how we should go about it. We may think about what we want to do, where we want to go, what we'd like to buy, who we'd like to meet up with – just about any aspect of our lives, really. But do we always do what we intend to do?

Not very often, it turns out. Partly, that's because human beings are gifted with imagination – the ability to visualize things which haven't happened yet, and perhaps, realistically, couldn't ever happen. And situations change all the time – we may find that we have other more important things to do, the weather might make it impossible for us to carry out our plans or we might simply not be able to afford whatever it is. Also, we change our minds – we have other ideas or realize that our first impressions were wrong and we need to think differently about something. So a lot of our plans change or turn out to be impractical when their time arrives.

But what about the plans that we really want to keep? Most of us make New Year resolutions – we decide to exercise more, lose weight, drink less or similar. Most people manage to keep those resolutions for about a week or so, but typically by 10 January they've been abandoned. No matter how good our intentions are, we simply don't stick with them for the long term.

But why is this? Well, a large part of the reason is that we expect too much of ourselves. As we've seen in previous chapters, we're not always rational, balanced people with a clear understanding of what we are like and what's going on. Some days we may be tired, angry or depressed, and simply not feel like putting in that extra effort. Or we may find that it takes up too much of our time, and it's unrealistic given our busy lifestyle. Or we may be unable to keep it up on one day and then feel that we have completely failed, so it's not worth trying any more.

The secret to keeping New Year resolutions is threefold. Firstly, we need to make sure that we have set **manageable goals**. It's just unrealistic for most of us to spend two hours exercising every day, or never, ever to eat chocolate. Setting a limited target that's easy to achieve – which means that it's easy to fit into our everyday lives – is much more practical, and massively increases the chance that we will keep our resolution.

Manageable goals: plans or intentions which are realistic and achievable given the availability of time, effort and situational options.

Which brings us to the second part of the secret: it's all about **habits**. What we really want to do when we're making a New Year resolution is to change our behaviour. But if that change is to become lasting, it needs to become a habit – something that we do unconsciously and automatically. And that can't happen if it involves a huge effort. It has to be (relatively) straightforward

Do you want to stop smoking?

Giving up smoking is particularly difficult for many people because the act of smoking has become so habitual. The activity is triggered by times of day, breaks in the pattern of work and, above all, our immediate setting. The trick is to replace smoking a cigarette with doing something else at those times. Some people find that vaping or chewing nicotine gum is a satisfying alternative, although it does have a high relapse rate; others find different options – there are any number of possibilities. One of the most successful strategies is simply to delay the first cigarette until a specified time – not so far into the day that it has become stressful, but just a little bit further than usual. For example, if the usual habit is to reach for a fag first thing, it is delayed until after the first coffee. If it's usually with the first coffee, it's postponed until after breakfast, and so on. The smoker keeps to the first target for at least two or three weeks, and maybe even a month, and then moves the target on just a bit, to the next focal point of the day. The target time for the first cigarette moves on very slowly, giving the new point enough time to become habitual. It's a slow process, but it's extremely effective, partly because it involves small steps that are easy to keep up, but mainly because it insinuates alternative habits into the day.

to do, so that it can be achieved regularly, and eventually without even thinking about it. If we've set manageable goals, the new behaviour should become easier and easier each time we do it, so that by 10 January it has become habitual and requires little effort at all.

Habit: a pattern of behaviour that has become so ingrained in everyday life that it is no longer controlled by the conscious brain, but occurs without intention or conscious thought.

The third and final part of the secret is to see the whole thing as a process, not as a sudden change. Obviously, a New Year resolution is something you need to keep up every day, as far as possible. But it's that last bit that matters. If you find that you have lapsed on a particular day, you need to do what the athletes do – pick yourself up and start again. There's no point throwing away a whole possible life-change because of one failure. Using disappointment as a way of justifying not bothering is a guaranteed way to fail. You need to accept your failure as having happened, refocus and persevere with the main effort.

20
ARE YOU ANGRY WITH ME?

We've all experienced occasions when we've been so angry that we did or said something that we wish we hadn't. It's not as if we stop thinking at such times. It's more that the type of thinking we do is completely different. Our emotional states have a powerful influence on our thinking. Being angry or upset channels our cognitions in many ways – to the point where it can even feel as if we have become a different person.

One of those ways is that our emotions affect how we interact with other people. We're all hard-wired to recognize the emotions of others. People in every human culture have been shown to recognize eight **basic emotions**: happiness, sadness, disgust, fear, anger, embarrassment, excitement (or surprise) and awe. Each of these involves different facial expressions and tones of voice. But although we might easily identify these emotions in others, we're often unaware of how we ourselves are communicating them. And other people react to how we present ourselves and behave, which directly affects the quality of our interactions. Nobody talks freely with people who they feel are angry with them, or even frightened of them.

Basic emotions: emotions that have been shown to occur and be readily recognized in every human culture, whether industrialized or not.

Our emotions don't just affect our behaviour, though. They also have an impact on how we interpret what other people are doing or saying. When we're anxious, we go into a mental state which makes us particularly vigilant to challenges – it's an ancient survival mechanism. So we're much more likely to interpret something that someone else says as a potential attack or criticism than we would do if we were calm. When we're angry, our mental state makes us particularly inclined to look for other sources of anger. It 'primes' our thinking so that we become easily irritated, and that affects how we react to new information or ideas.

Emotions influence our thinking in other ways too. They directly affect what we remember, as well as the possibilities we can imagine. Most people in long-term relationships have had direct experience of this: when someone is angry, all they seem to be able to remember are the other times when their partner made them cross; when they are happy, they remember the wonderful moments that they and their partner have shared. We have so many memories to draw on, particularly in a long-term relationship, that our brains automatically select the ones that seem most relevant to the state we are in. This can seriously distort what we say and how we interact with other people, because we forget about the other times and how important they were.

It isn't just about extremes like anger, though. Our moods have a powerful effect on the way that we think in general. It may seem obvious, but we think more clearly about problems when they are relatively easy to reflect upon – when we can identify their important aspects and then work out a solution. That's more likely to happen when we are calm and relaxed.

Helping with bereavement

It's the way that extreme emotions disturb our thinking that makes the support of friends and family so important in times of grief. People who have experienced the death of a partner or close family member are often numb and bewildered at first, as their brains try to absorb such a huge loss in their lives. Thinking about anything else becomes impossible. At these times, the best support from neighbours and friends is often to avoid asking the person to think about anything, but just to deal with the simple, practical things that wouldn't normally be any trouble at all. Making a cup of tea or putting a meal in front of the bereaved person, helping them to find documents or fill in forms, even tidying up the kitchen – all of these simple actions can make a lot of difference to someone who is grieving because their loss is so overwhelming that they feel completely unable to think at all. It passes, of course, as they gradually regain control of their lives, but while it lasts, that sort of help is really worthwhile.

Psychologist Daniel Kahneman refers to this as **cognitive ease** – and it means that most of our thinking isn't particularly challenging and will only be straightforward, system 1 thinking. Unless, that is, we are angry, upset or grieving, when even

Cognitive ease: a state of being relaxed, contented and feeling competent, such that thinking is experienced as straightforward and unproblematic.

everyday things can become difficult to think about. Being uncomfortable or unhappy directly interferes with the way that we think – it affects how well we deal with everyday challenges, how we go about working things out and whether we just give up on them altogether, how well we do in exams or tests and so on. In short, it affects our thinking. As all good teachers know, we learn best when we are relaxed and happy: we think best at those times too.

4

PERCEIVING

In this part of the book, we will be looking at how we use and make sense of the information we receive through our senses. For human beings, sight and sound are the main senses that keep us in touch with the outside world. It's striking, for instance, how people can attract our attention with a single word and how we respond almost automatically to particular sounds. It's equally fascinating how the brain analyses visual information, like the way it has special areas for recognizing faces and for identifying familiar people from their body movements.

We have many more senses than we might think: humans have at least thirty-two senses at the last count. They range from our ability to detect temperature (thermoreception) to the sense of how our bodies are positioned at any particular time (proprioception). But extrasensory perception – that which doesn't come through any of our known senses – is another matter.

21
DID SOMEONE SAY MY NAME?

One of the characteristics of the system 2 thinking that we explored at the start of the book is the way that it needs our full attention. System 1 thinking is pretty much automatic, but system 2 thinking demands much more of our 'cognitive space'. Psychologists have been studying the concept of attention for more than a century and been continually intrigued by some of its characteristics.

One of these is the way that we can only really pay attention to one thing at a time. We can switch between different activities quite rapidly – you might simultaneously be following a drama on TV while playing a (relatively simple) computer game with one hand and eating a pizza with the other. But at any one moment, your attention is on only one of those actions and, if someone in the room speaks to you, on none of them. Attention is like a kind of searchlight for our awareness – when one thing is lit up, the others blur into the background.

But not completely. If someone says our name, for example, our attention immediately switches to the speaker. If we were totally focused on what we were doing to the exclusion of everything else, we wouldn't even hear that, so we must have been aware of the sound at some level for it to have attracted our attention. Psychologists have devoted much research over many years into developing models of **selective attention,** which can help to explain how our brains do that. It turns out to be quite a complex process, but essentially there are specialized neurones in our brains that respond to key personal information like our own names. Activating them seems to be able to override the neural processes involved in concentrating.

Selective attention: attention that is directed towards a certain source of incoming information and to nothing else.

Attention can be a strong channel for our brain activity, tuning out most external stimuli. But even when we are focusing closely on something, we can be suddenly distracted. We are biologically prepared to notice sudden sounds and also pain. Each of these sets off a series of **physiological responses** that heighten our receptiveness to relevant information and demand our conscious attention. They are ancient survival mechanisms and, as anyone suffering from chronic pain will tell you, exceptionally difficult to ignore. Extreme emotions like grief or rage have a similar effect on our thinking: they really interfere with our ability to pay attention to other things around us, and so does worrying. In all of these cases, our conscious minds are taken up with their demands, making it harder – not impossible, but more difficult – to notice what else is going on around us.

Have we been brutalized?

Words that are negatively emotive – that is, which arouse negative feelings in us, such as war, torture or death, tend to attract our attention much more than happy words do. This is likely to be another remnant from our primeval past, in that staying alive and healthy, for the most part, will have involved avoiding such things, so hearing other people mention them would have been really significant. Interestingly, we can damp that sensitivity down with enough exposure. Research has found that people in industrialized societies who are exposed to a lot of aggressive and violent film and TV dramas react much less strongly to negatively emotive words and images of violence than those from other cultures or people from their own culture who don't watch much of that type of entertainment. There is some anxiety (and a bit of psychological evidence) that it may carry over into real life: that some individuals have become brutalized by that vicarious experience to the point at which they have become insensitive to real-world torture or killings. It's a contentious issue, with massive economic and social implications for the entertainment industry, so the debate continues.

Physiological responses: complex bodily reactions which involve a combination of changes in different physical processes.

22

WHO'S THAT OVER THERE?

Our brains are amazing. They can put together all sorts of diverse types of information and produce results that we can actually use in our everyday lives. As anyone working with robotics will tell you, that's not an easy thing to replicate.

Take the way that we see movement. Our brains automatically link things together: if you show someone a set of dots on a screen and you light them all up together, they see a series of dots. But if you light them up one at a time, with each lit dot going off as the one next to it comes on, people don't see a set of dots. They see one dot moving across the screen. It's called the **phi phenomenon** and it's one of the earliest observations made by psychologists studying vision. More to the point, it's the basis for the entire film, TV and video-game industry.

What any movie is really showing you is a series of still images. But they are presented so quickly that your brain automatically registers them as continuous movement.

Phi phenomenon: an illusion of movement brought about by a rapid succession of lights or images.

Have you ever had the experience of seeing a friend a long way away – so far that you can't make out anything much except a blurry stick figure? It doesn't take you long before you recognize who it is. And that's remarkable. What your brain has done is take some absolutely minimal visual information, identify that it's a person, recognize its overall shape and the pattern of its movements, and link it with your personal knowledge of other people. Of course, sometimes we make mistakes and it turns out not to be that person at all (which can be embarrassing), but we're right more often than we are wrong, and the fact that we can do it at all is astounding.

But how do we even know that the blurry shape we see in the distance is a person at all? The neuropsychologist David Marr developed a theory of **computational perception**, which showed how we can use the most basic information – even just the dots of light and dark that your eyes might receive for distant

Computational perception: the way that the brain analyses boundaries and edges in our vision and applies mental algorithms to work out what they represent.

objects – and combine it into simple shapes, like stick figures. Then we use a stored mental 'catalogue' of typical shapes to begin to identify what we are looking at.

A dog or cat, for instance, makes a different stick-figure shape to that of a human being. We can combine an impression like that with other information about distance to get an idea

Did you move?

We're particularly good at recognizing other human beings. You may have seen a demonstration of a dark space with what seem to be random dots of light. Suddenly the dots begin to move, and instantly you stop seeing random dots. Instead, you see people moving around. We don't just see the movement; we instantly make sense of it, to the point where we could almost sketch in the human figures doing the movements. These demos are made by people in black bodysuits on a black background, with lights attached to them in various places. They work for animals as well – we can easily recognize dogs, cats, even (memorably) elephants when they start to move. It's less effective for machinery, though, because machine movement tends to be more linear and regular, and also because machines don't really form part of our evolutionary past in the same way that animals do. So we're not hard-wired to recognize them.

of its size. So we might mistake a cat for a dog, but we wouldn't confuse a cat with a person or a horse. When it comes to identifying people, there aren't many other things that the stick figure for a human could be, and the pattern of movement helps too. So our brains are able to compute what they are looking at and draw on stored knowledge to make a firmer identification.

When it comes to closer contact, we have specialized areas of our brains that tell us who we are looking at, such as one that detects facial features – eyes, nose, mouth and so on. Even very tiny babies have been shown to smile at images that have two dots, a curve and a line arranged into a face-like shape. The range of smileys that have become part of our library of emojis shows how much information we can deduce from even the most basic details.

But that's not all. We have two other specialized regions of the brain that help us to recognize faces. One of them reacts particularly to the parts of the face that we use when we're communicating – lips, facial expressions, gaze and so on. That area of the brain is also involved in lip-reading and it has direct connections with the parts involved in hearing speech. Which is why wearing a mask makes it so difficult for other people to make out what you are saying.

The third part of the brain that's connected to facial recognition is known as the fusiform face area, and is the bit we use when we are identifying who someone is. It's close to the memory parts of the brain – not surprisingly – and also to the regions concerned with emotions. That's why we can experience an emotional reaction to someone's face even before we've realized who that person is, and why we do or don't like them.

23
YOU'RE NOT MY WIFE!

Other people are so important to us that recognizing faces and body movement is hard-wired into our brains. So what happens if that recognition doesn't work properly? As we saw in the last chapter, we have special areas of the brain for recognizing faces and also for identifying people we know well. That's probably one of the most important skills we have – our lives pretty well depend on being able to interact with other people, so it's a very deep part of how we think. That depth is literal as well as metaphorical – the areas of the brain which do this are buried really deeply inside the **cerebral cortex**: they are on the part which folds round and tucks right underneath the main part of the brain. So there's much less chance of them being accidentally damaged through a head injury or accident.

Cerebral cortex: the outer part of the brain, which is folded and grooved in human beings and covers over almost all of the rest of the brain.

Those areas can fail in other ways, though. Diseases like dementia can cause degeneration of the brain cells involved, or minor strokes or other disturbances can result in those parts of the brain not functioning in the right way. And the outcomes

Who's that in my house?

The brain disorder known as Capgras syndrome causes people to become convinced that their close friends and family have been replaced by strangers who are impersonating them. As you can imagine, both their loved ones and the person concerned find that highly distressing, but it's extremely difficult to convince the sufferer that it isn't true. That problem appears to arise from the fusiform face area – the part of the brain that identifies faces – becoming disconnected from the regions of the brain that process emotion. Because our recognition of close friends and family is always accompanied by some emotional response – that's part of why we feel that they are close to us, after all – seeing their objective features without that emotional link naturally makes a person with Capgras syndrome feel that they aren't the real individuals.

can be dramatic. At the least, people affected by those types of health issue may just become unable to recognize people who are relatively unfamiliar to them. But if other areas of the brain are also damaged, they may no longer be capable of recognizing close friends or even loved partners. As anyone with a relative suffering from dementia will confirm, that can be extremely painful to witness.

Neurologist Oliver Sacks called his famous book of case studies of brain damage *The Man Who Mistook His Wife for a Hat*, and in it he described exactly that scenario. The man concerned suffered from visual agnosia, which means he couldn't identify what he was looking at. Seeing a hat on a peg, he assumed it was a person, namely his wife. There are more common examples too. People with dementia, for example, are sometimes able to recognize the face of the person they are talking to, but they misidentify it, so they think that the individual is someone else – often from their distant past.

Recognizing faces is an important social skill. But some individuals, who may be perfectly normal in all other respects, find it difficult or even impossible. They have what is known as **prosopagnosia**, also known as face blindness. People with prosopagnosia may recognize others' individual traits, know all about them and even be able to identify them from the way that they move, but they cannot work out who they are from seeing just their faces. For some people, face recognition simply doesn't work particularly well and nobody really knows why. In one case, for example, a man was entirely unable to recognize faces, even that of his own wife. But he had a massive collection of miniature cars and could distinguish between every single one of those.

Prosopagnosia: a brain disorder which produces an inability to recognize faces.

There are degrees of prosopagnosia, of course. Most of the people who have it are perfectly well able to recognize friends, family and those they have known for a long time or see often. But they don't recognize people who are less close – those who they may have met at social gatherings or who are from other departments at work. A surprising number of people have this kind of prosopagnosia – even some very famous ones – but they are usually able to cover it up, perhaps by talking to the person and getting clues about who they are, or using some other social strategy.

24
WHAT TRIANGLE?

Our thinking is primed to make sense of what we see and hear. We do this so easily that most of the time we're not really aware of it. But the world around us gives us a lot of information and there are several cognitive mechanisms we use, unconsciously, to analyse it.

For example, the image our eyes receive gives us cues to distance – that is, it provides hints as to how far away things probably are. Things that are further away appear higher up in our vision, and if they are very far away, they seem greyed out as well. They also look smoother because we don't see their detail and they may be partly covered up by nearer objects. All of these – height, greyness, texture, being partly covered – are **depth cues**, which let us know what is near to us and what is further away. They are the visual hints that someone might use in a painting, though that's a still image. The fact that we move around in our world means that we can use other cues as well, such as motion parallax, where things seem to move differently depending on how far away they are and how we look at them. You can try this by holding out a pencil and lining it up with something nearby. If you move your head from side to side, the pencil seems to move backwards and forwards relative to the farther object.

Depth cues: those features of a visual image which imply how far away something is.

As a general rule, cues like these and a few others are all that we need to tell us where things are, whether they are coming towards us and if we might collide with them. Our brains do the sophisticated calculations that help us to be active in our environments without our even being aware of it. But sometimes, our brains can be tricked into misinterpreting what they see.

We're programmed to apply depth cues to what we see too, even with minimal information. The Ponzo illusion consists of two nearly vertical but slanting lines, closer together at the top than the bottom. If two equal-sized horizontal lines are placed inside them, one above the other, the one at the top seems longer. That's because the brain interprets the slanting lines as

Illusions in art

Many artists have incorporated depth cues and illusions in their work. The artist M. C. Escher was great friends with many of the Gestalt psychologists who scientifically identified the principles of perception, and he built many of their ideas into his artworks. One of the principles they identified was the way that we always see figures against backgrounds, but never the two together. Escher did many illustrations using this idea, with images consisting of two figures where each forms the background to the other. There's a famous one of angels and demons: if you look at the demons, the angels disappear, but when you look at the angels, the demons simply become the background. You can see them both, but not at the same time. Other Escher artworks use perspective illusions, like the steps that look as though they are always heading upwards as they go round a square tower. Check out Escher's art and you'll find lots of examples like this.

if they were pointing into the distance, like a path or roadway. So it thinks that the one higher up must be further away. If it was the same size but further away, it would look shorter. Therefore it must be bigger.

This type of error is common in a lot of visual illusions. They are known as strategy illusions because they come from the cognitive strategies that the brain uses to make sense of incoming information. Other kinds of illusion include mechanism illusions, which use the physiological mechanisms of the visual system, like the phi phenomenon we looked at in chapter 22, which creates an impression of moving images.

Another common mechanism illusion is the waterfall effect, which comes from **habituation** – like the way that you don't notice the humming of a fridge until it suddenly stops. If you stare for long enough at a waterfall, when you look away you feel as if what you are looking at is moving upwards.

Habituation: the way that nerve cells stop responding if they are exposed to repeated identical stimulation.

Similarly, if you gaze out of a train window for any length of time, when the train stops you might have a visual impression that things are moving backwards. These are negative after-effects: your visual cells have become so habituated to that stimulus that you get the opposite effect when it stops. It works with colours too – stare at a red jumper for long enough and when you look away you'll see a bluey-green jumper shape in your vision.

25
I WASN'T EXPECTING THAT!

An awful lot of our perception is unconscious – although it's an important part of how we think, it happens at an automatic level so we're not really aware of it. But as we move around the world, we are constantly drawing on our stored knowledge to make sense of what is around us, and also to tell us what we should be expecting next.

Have you ever been walking along a road, and suddenly stopped because you've seen a particularly odd sight? It might be a peculiar vehicle, like an old-fashioned steamroller, or something really strange in a window. It attracted your attention because it was so very different from what you were unconsciously expecting to see. And that tells us something else about the way that we think.

In chapter 2 we looked at mental set – the condition of being prepared to think in certain ways, which can mean we don't even realize that other ways might be better. We have **perceptual set** as well, which relates to how we can be particularly ready to see or notice some things rather than others. As a general rule, we know roughly what to expect at any one time: it's very rare that we are in a situation where we really don't have any idea what might happen. So we are pretty well prepared to come across the usual sights associated with everyday living – it's only unusual things that stop us in our tracks.

Our general expectations don't just tell us how to make sense of things: they also act as schemas: stores of knowledge that we can use to guide and direct our actions. We'll be looking at schemas more closely in chapter 28: they are more than just memories because they contain knowledge about what would be appropriate actions as well. So, they are particularly important when we are actively moving around the world because they are a guide to how we should act and what we should notice.

Perceptual set: a state of preparedness to perceive certain types of information rather than others.

But if we only followed our expectations, we'd never notice anything unusual – and indeed, as we've already seen, sometimes we just don't. For the most part, though, our everyday thinking follows a continuous cycle, with each part leading on to the next. It was identified by Ulric Neisser and it's generally known as the perceptual cycle.

As we walk along a road, for example, we are using what Neisser called an **anticipatory schema** to give us a rough awareness of what we can reasonably expect to happen next. That schema tells our perceptual system what it should be focusing on, and our perceptual system samples the real-world information available: even on a quiet day, our senses receive far too much information for us to notice everything.

The cycle part is this: we're constantly sampling the real world through our senses. The information that is sampled informs the anticipatory schema, which is then adjusted or modified to take

Where have the baked beans gone?

One of the more irritating experiences of day-to-day living is popping into a familiar supermarket and finding that the layout has been changed. Everything has been moved around and you have to search for the things you want to buy instead of going straight to the right shelf. It may be irritating, but it's not accidental. From the moment we enter the store, our anticipatory schemas are in action: we know exactly which direction to go in if we want to find washing powder or baked beans. But this familiarity also produces the inattentional blindness that we looked at in chapter 2: other things simply don't attract our attention. Once we're completely accustomed to the layout, we might not even notice a really striking display – much to the frustration of the store manager! If it's all been changed around, though, we have to take notice – in Neisser's terms, we have to sample the environment much more and modify our anticipatory schemas accordingly. So we're much more likely to notice new things. There's a caveat, though: if it happens too often, we get frustrated and just focus on completing our shopping lists without looking around. Novelty only works when it's novel!

account of it. That new version of the anticipatory schema then directs our perceptual focus – what we pay attention to – which samples more real-world information that then modifies the anticipatory schema, and so on. It's a continual process: we may not be aware of it but we're doing it all the time, and it's the core of how we notice (or don't notice) what is going on around us.

Anticipatory schema: a mental representation of our surroundings, which tells us what to expect and is continually updated as we move.

You can see, then, why you might suddenly stop in the street when you see that steamroller or the particularly dramatic shop window (imagine how delighted the window-dresser would be!). Before that, you were hardly noticing the traffic or other pedestrians, although you were behaving appropriately in response to them, and even unconsciously taking evasive action to avoid bumping into people. But the steamroller or shop window attracted your attention because they were so entirely different from what your anticipatory schema had predicted.

26
DO YOU REALLY KNOW WHAT I'M THINKING?

Many people believe that there are aspects of life of which we are completely unaware, like telepathy or telekinesis. Many also believe in phenomena like ghosts, the Loch Ness Monster or UFOs, but that isn't the kind of thing that the psychologists who study **parapsychology** are interested in. Parapsychologists are primarily concerned with the abilities and experiences of living human beings, and they are scientists dedicated to gathering evidence, so while they are interested in telepathy, telekinesis and other possible abilities, whether Nessie actually exists is outside their focus of interest.

Parapsychology: the study of '**psi**': the human ability to detect or transfer information without the use of the known human senses.

The problem is that there is a long-standing and extremely clever tradition of fooling the public with demonstrations that appear to show ESP (extrasensory perception), but are actually very sophisticated magic tricks. This means that serious parapsychologists who want to investigate these abilities have to be particularly careful in how they go about their research. There are four main areas of parapsychological research: telepathy, clairvoyance, precognition and psychokinesis.

Telepathy is communicating information from one person to another without that information using the known external senses, such as olfaction, sight, hearing and thermoreception. The information concerned might be conscious thoughts, but more often it's vaguer, like having a strong feeling that something has happened to someone and then finding out that they've just had an accident. The problem is that it's very difficult to study that sort of thing objectively. We can't evaluate it simply from people's reports because there are so many other factors that can influence what they remember.

Psi: a general term used to describe mental abilities that can't be explained in conventional ways.

Clairvoyance is obtaining information about a place or event by unknown means. In the 1970s, there were several studies of remote viewing, which seemed to suggest that some people could travel mentally to faraway places and identify buildings or other features. Understandably, this research attracted considerable interest (and funding) from the defence industries, but in the end the outcomes were too vague to be of real interest.

Precognition is another problematic area. Although there may well have been some instances where people seemed to know in advance that something would happen, actually pinning those cases down for scientific scrutiny is extremely difficult. And it isn't made any easier by the way that common errors of thinking can distort what we remember. **Hindsight bias** (also known as the 'I knew it all along' effect), for example, can mean that we believe we knew something before we really did; and studies of eyewitness testimony show how our memories of events change with our assumptions and expectations. We'll look at that more closely in chapter 39.

Hindsight bias: the tendency to see things as having been obvious when looking back, even though they were far from obvious at the time.

In some cases, too, there's the statistical argument. House fires, for example, are a source of worry for many people and it's pretty common to dream about them. On any given night, there may be thousands of people having such dreams. Statistically, it's actually quite likely that one of those people may then encounter a real house fire. Was their dream precognitive? It might feel as though it was to them, but most scientists would be more sceptical. A one-in-a-million event can happen sixty times in a population of 60 million. And most people have several dreams in a single night, even if they don't remember them.

Psychokinesis, or telekinesis – the ability to move objects by mind power alone – is a clearer challenge. But, unfortunately, it's also the area in which stage magicians have had the most practice – and they've become extremely good at it. The magician

Mystical or mechanical?

In the 1970s, there was a wave of TV and media interest in spoon-bending and other apparently psychokinetic experiences. Many of the demonstrations were highly convincing to the public. On one show, for example, people were invited to find old watches that didn't work any more and hold them tightly in their hands as the 'psychic' extended his influence through the TV screen to make them work again. There was great excitement as viewers across the country found their old watches ticking away again, which led to a huge growth of belief in 'psychic' influences. The problem, though, is that these were all old wind-up watches, and the usual reason why clockwork of that kind fails is because a tiny bit of dirt has got into the mechanism. As the watch was being held tightly in its owner's hand, it warmed up, so the lubricating oil became more fluid and able to move the speck of dirt. The effect was simply mechanical – but appeared 'mystical' to the owners. And, of course, we simply didn't hear about all the watches that didn't restart.

James Randi offered a million-dollar prize to anyone who could conclusively demonstrate telekinesis on real objects; but he's been able to show how all of the attempts to win the prize were actually very clever deceptions. The same goes for many apparently convincing laboratory investigations in the past. So modern lab research into psychokinesis has to be rigorously controlled to prevent deception from con artists and tricksters.

Parapsychologists continue to investigate psi, or extrasensory, abilities, but in view of the many past disappointments, and knowing how the media often exaggerate scientific findings, they are understandably reluctant to announce positive results, even if they think they may have found them. They continue to test and re-test until they can be absolutely sure. The standards they have to keep to are far stricter than those for just about any other form of science because there are always sceptics who are ready to challenge them. So any suggestion of 'leakage' of information, no matter how unlikely, would be taken as invalidating their findings.

5

REPRESENTING

As we go through life, we accumulate vast amounts of knowledge, from how to boil a kettle or knowing what a giraffe looks like to being aware of where to go to see the Northern Lights. We have to organize all this information somehow and we do it in many and varied ways. We use concepts for some things, sensory impressions for others, and we even develop our own personal theories about other people and what they are like. This section of the book is all about how we represent information in our minds.

We represent social information too, like how people say things. For example, we might judge someone's personality differently if they have a 'posh' accent, by comparison with someone with a strong regional accent, even if they are actually using the same words. We have definite expectations about appropriate behaviour in social situations. And when we really feel out of control, we fall back on superstitious representations about what will bring us luck.

27

WHAT ON EARTH IS THAT?

It comes very naturally to us to categorize things. We automatically sort what we encounter into groups – types of people, varieties of animals, kinds of environments, makes of cars – the list is endless. It's actually fundamental to being able to interact effectively with the outside world. If we had to treat every different plant, building or person as if they were totally independent and unlike anything or anyone else we had ever met, our brains would be completely overwhelmed – they could never process all that information.

So we develop **concepts** – categories that are based on what things have in common. These concepts operate at different levels – some are very general, such as people, while others are more specific, like BMW drivers. At the most basic level, we might have categories like animals, plants, people and so on. Slightly higher up we have types of animals, plants, etc. – such as the difference between domestic animals, farm animals and wild animals, or between flowers, bushes and trees. We add levels as we get to know more about an area – for example, people who are not gardeners may simply lump all types of flowers together, while a keen gardener would distinguish between all sorts of varieties.

Concept: a general term that we use for a group or category of objects, events or ideas.

There have been many attempts to describe how concepts work in the mind. Some psychologists argued that they were all about features – a chair is something which (usually) has four legs and a back, for instance, while a stool may have fewer legs and doesn't have a back. The fact that there are often variations in these features led to the idea that the concept actually reflects a basic **prototype** or a shared central idea for a concept. Others, though, have argued that concepts represent natural categories based on how we actually use or interact with them. For example, the category 'furniture' is all to do with things that facilitate our everyday life – that is, actions, in that the subcategory 'chair' is something we sit on, while a 'bed' is a thing we use to sleep on and a table is an item we put things on.

Prototype: a model of a typical example of a particular concept, which contains all of its essential features.

Concepts are useful as a model of how the brain represents physical objects, but we have other forms of representation too. Words, for instance, are not physical; for human beings, they are fundamental to categorizing as well as sharing information about our worlds. Developing word recognition is key to fluent reading – good readers recognize words quickly, by their shapes, and associate them with other related information very rapidly.

We can see how rapid that extra association is by the Stroop effect. If you show someone a list of colour names printed in their own colours, they can read them out very quickly. But if the names are printed in different colours – for example, if the word 'red' is printed in blue and 'orange' appears in purple type, the person takes much, much longer to read them out, and may even make mistakes. You can do the same sort of test with words written in UPPER CASE, **bold**, *italic* or lower case. The Stroop effect shows us that word recognition isn't just about identifying the words themselves – it's also about representing their meanings and associations.

We can represent things spatially, as well. As we get to know different places, we develop cognitive maps that tell us where things are in relation to one another. These maps are very personal and they change with our experience. You may remember, for example, that when you first went to a new place, like a university or a town, the distances between different areas seemed much larger than they were when you got used to it. The way that we represent such places is closely tied up with how well we know them. Even animals can develop cognitive maps – for any mobile animal it's important to know roughly whereabouts in its world it happens to be, and human beings are no exception.

Medieval map-making

One of the fascinations of looking at medieval maps is the way that they show, so clearly, how the mapmaker saw the world. They are, effectively, cognitive maps drawn on parchment. Those ancient maps, generally drawn up by Christian monks, feature the centre of the world as the centre of the Christian religion – that is, Jerusalem. The rest of the map is in as much detail as the mapmaker knew, so at the edges are unknown areas, often populated by monsters – 'here be dragons' – while European areas are represented in considerable detail. Some equally ancient Chinese maps depict roads and routes from one place to another: places near the road are shown in detail, while areas that are further away are only vaguely suggested.

Distortions are not only true of ancient maps, though. The standard Mercator map exaggerates the size of countries located away from the equator, like Greenland, and reduces the size of those in the centre, for example making Africa appear smaller. The Gall–Peters projection gives a more accurate country size, but distorts the distances between them. This generated considerable social debate as the Mercator projection made rich countries appear bigger: another example of cognitive mapping, in that the shape and size of the countries represented how different groups of people perceived them.

28
HAVE I SEEN THAT BEFORE?

We don't just store knowledge, we use it. And the way that we represent information in our minds is closely tied in with how we do that – how it fits with our understanding of possibilities and actions. The way that we do this is through **schemas** – units of representation that include concepts but also feature wider knowledge, such as potential actions, associations and experience. Mental schemas are the way that we connect knowledge with actions, and memory for actions. New experiences are absorbed into schemas and adjust these schemas at the same time.

Schema: a hypothetical mental structure that includes plans and actions relating to an idea as well as its relevant information.

The original schemas begin to develop with our very first experiences of the world. When an infant is born, its senses are bombarded with information from the outside world. Gradually, it learns to differentiate between 'me' and 'not-me' – and those are the first two schemas we all encounter. As the young child's physical control and its senses become more sophisticated, its schemas do too – for example, babies are particularly inclined to respond to other people, so 'person versus non-person' is likely to emerge as a schema fairly soon.

Schemas grow by assimilating new information – by being applied to new situations. Sometimes that doesn't involve change, but occasionally the schema has to adjust itself, stretching and growing to be able to encompass new information. That's known as accommodation. If the schema becomes too stretched, it may separate: a schema that begins as 'non-person' in a baby's world may become two schemas: one for cuddly things and one for hard things, or food and non-food (although, as parents know, that bit of learning usually comes much later!).

Egocentricity: not selfishness, but the sense of being the entire centre of one's world, with everything else only having relevance as it impacts upon it.

One of the major developmental psychologists of the early twentieth century regarded schema development as being the key to understanding the child's cognitive development. In particular, Jean Piaget said that we all start off as totally **egocentric** – a baby is the core and centre of its universe. Learning to deal with the world involves gradually reducing that egocentricity, as we become aware that parts of the world are separate from us and independent of our own experience. You can understand this when you observe small children hiding: often, a child will cover its eyes so it can't see you and then assume that you can't see it either. It takes some time before very young children realize that other people can see things even if they can't.

Piaget wasn't the first person to describe schemas. A couple of decades earlier, another psychologist, Frederic Bartlett, reported

Does it fit your world view?

When we encounter new ideas or experiences, we need to find ways of incorporating those experiences into our schemas, or of adapting our schemas so that they can deal with them. If we can't do that, we often just end up rejecting those ideas altogether. There are people, for example, who won't read the Harry Potter books because they believe they are only for children, or some who find it almost impossible to enjoy science fiction because they are unable to suspend their real-world schemas and accept an idea that is fundamentally impossible, such as faster-than-light travel, which has to be taken for granted in the book or film. (Interestingly, those same people may have no trouble with a TV series in which murders happen in the same village practically every week, which is equally implausible really.) Those who have more flexible schema structures, however, can deal with the premise of an imaginary universe much more easily.

a famous study in which he showed how memories become distorted as individuals struggle to fit them into the ways that they understand the world. What he did was to ask some people to listen to a story and then recount the same account to another person, who would then share it with someone else, and so on. The story was widely outside of their own experience – it came from Native American culture and involved the active intervention of the spirits in real life. As those people passed it on, the content changed to become more conventional – to fit into their existing schemas better. It wasn't deliberate – they didn't know what they were doing. But as they re-told the story, they only remembered the details that could be fitted into their existing schemas.

Our schemas affect what we notice as well as what we remember. In one study, some people were asked to watch a video featuring footage taken in the rooms of a house. One set of viewers was told to imagine that they were prospective house-buyers; the others were instructed to take the point of view of a potential burglar. When they were asked about the rooms later, the two groups remembered very different details. The schemas they had been using had affected the entire way that they looked at the house.

29
IN MY MIND'S EYE

We have seen how representation is more than just storing factual information. We don't just record information like a camera: we are personally involved in the ways that we represent our worlds. Our ideas about what the world is like draw on our own experiences and ideas, as well as on external information.

How we go about representing information also draws on the sensory input that we may be feeling. Most of us have experienced a wave of memories rushing over us when we encountered a distinctive smell, or heard a particular song or tune that has taken us back to another time and place. That sensory information has become closely linked with our personal, autobiographical memories.

The very first form of representation we develop is known as **enactive representation**. This is muscle memory, where things are represented in terms of the actions we carry out. You can see this in babies, who may display a particular action to indicate, for example, remembering a rattle or squeaky toy; but we also retain this ability as adults. Think of the experience of going

Enactive representation: the kind of 'muscle memory' where the person experiences a mental impression of what an action would feel like.

on a Waltzer at a fairground, or of travelling round a corner at high speed as a passenger in a car. You'll probably remember how your body felt as the g-force hit you – that's enactive representation.

As the child becomes more sophisticated in how it interacts with the world, enactive representation is no longer enough. Some things, like reading a book or watching TV, involve different input but the same muscle actions. So other forms of representation develop, and one of the most important is known as **iconic representation**. An icon is a small picture, so this is representation using pictures or visual images.

Iconic representation: impressions of ideas or memories that are like mental pictures or images.

Many children have a particularly strong form of iconic representation which gives them an eidetic memory or photographic recall. They can explore the details of something they have seen just by remembering it, even if they didn't particularly notice those details at the time they happened. Roughly 10 per cent of children under ten years old can do this, but as we get older we don't retain the skill – less than 1 per cent of adults have eidetic memory. That's partly because we come to use other types of representation instead, so iconic representation becomes less important to us.

We need to develop other types of representation as our understanding of our worlds becomes ever more complex.

Are flavours coloured?

Some people use sensory impressions to represent information in ways we might not expect. There is a sensory disorder known as synaesthesia, in which sensory impressions appear to have become jumbled up, so sounds may be experienced as lights, while tastes come across as colours. Those are just examples – synaesthesia manifests itself differently in every person who has it. True synaesthesia is quite rare, but synaesthetic imagery used in representation is much more common. Someone might represent savoury flavours as 'brown', for instance, or describe (to themselves) the touch of velvet as being like the sound of a glockenspiel. Neurologists have yet to find any particular 'rules' to synaesthesia or to synaesthetic representation: it isn't a simple question of neural pathways crossing over or anything like that. Like memory and ability, it's different for each of us: how we represent knowledge in our own minds is a deeply personal and individual matter.

Ideas like justice or freedom, for example, are abstract and don't lend themselves to clear visual images. The same applies to mathematical and arithmetic skills: we can deal with that information easily using symbolic representation, but it would be much harder to do it using iconic imagery. Symbolic representation is basically what it sounds like – using symbols to represent knowledge. In a sense, we're using symbolic representation from the time we learn a language because words are symbols indicating a common meaning. But the more we move towards adulthood, the more important the various types of symbolic representation become, until eventually enactive and iconic representations take second place and most of our representation involves symbols of one kind or another.

Symbols are large abstract representations of ideas, but as we've seen, the way we represent information also includes our personal involvement with it. That involvement might be planned actions, emotional connections, how something has become associated with a specific person, an episode that we particularly noticed in a TV drama or film, or just about any other aspect of our experience. So we also develop semantic representation, which isn't just about storing facts, but about meanings. It includes moods and feelings like warmth or empathy, emotions such as anger or happiness, and many other aspects of our personal lives.

30
ARE REDHEADS HOT-TEMPERED?

People, in one way or another, are the most important things in our lives – and that goes for all human societies, technological or otherwise. So it shouldn't be surprising that the way we represent information about other people is a little bit different from how we represent other types of information, like the colour of grass or how far it is to the shops.

It's also more powerful: as a general rule, we take much more notice of things that we hear from other people than we do of other types of information. It sticks in our memories better. That's why students find it helpful to work in revision groups, quizzing each other about the facts they are trying to memorize, sharing and comparing summaries, and so on. As long as they manage to keep to revision and avoid getting sidetracked by chat or music, it's by far the most effective use of their study time.

This tendency to take more notice of what other people tell us can also lead us into one of the more common errors in everyday logic: the way that we are much more likely to believe or draw conclusions from a single example than from any number of statistics. Personal anecdotes are much more memorable: if someone you know, or a friend of a friend, had their wallet stolen while visiting a particular town, you would be likely to see that town as a significant location for theft crime – even if the statistics showed that such incidents are actually more unlikely in that place than elsewhere.

In modern life, we meet strangers all the time. Apart from being a slight source of unconscious stress, because we are just that little more tense in the presence of people we don't know than we are with friends, it also means that we are bombarded with masses of information about other people, which we have to represent somehow in our minds. We do this in two ways: partly by applying **implicit personality theories** and partly by using the personal constructs that we have developed

Identifying your own personal constructs

There's a pretty easy exercise you can do to find out what your own personal constructs are. The first step is to identify eight people who are important to you in some way. You don't have to like them, they just have to matter to you somehow. Allocate each person a letter between A and H. Then take them in sets of three and think about how two of them are similar to each other but different to the third. Write down those ways. This will give you a pair of words. Then take another triad and do the same with that. That will probably give you a different pair of words. Go down the list, including everyone at least once, but ideally two or three times. Here's a list of triads to help: ABC, DEF, AFG, BDH, CEG, HBF, AEH, DGC. By the time you've done this a few times, you might find some repeats, but you'll have ended up with a list of pairs of opposite characteristics, like 'quiet–noisy', or 'thoughtful–impulsive', or even 'rich–poor'. Those are your personal constructs. If you can get a friend to do this as well, you may find some interesting differences.

through our own experience.

Society is full of implicit personality theories. The idea that people with red hair are likely to be hot-tempered is one, as is the idea that fat people are jolly or librarians are quiet,

Implicit personality theory: unconscious but well-developed ideas about the personality characteristics that different types of people are likely to have.

mousy people. Studies have shown how simply describing someone as a company manager or a trade unionist is enough to generate theories about what type of person they are and how they are likely to act. In fact, we are able to construct whole representations of people – their habits, homes, cars, all sorts of things – based on the flimsiest bits of information. Moreover, researchers have shown that in constructing these ideas we completely ignore statistical information, which shows that the persona we have created is highly unlikely.

The second way that we represent other people is through the **personal constructs** we have developed in our lives. Personal constructs are small mini-theories about what people – and other things in the world – are like. They take the form of bipolar constructs – that is, dimensions which have two opposite ends, like 'kind–cruel' or 'passive–active'.

Personal constructs: individual ways of making sense of the world, developed through personal experience.

We each develop our own unique set of constructs, and people typically use about eight or so of them in making sense of others. Be warned, though: the words we use to describe our constructs may be similar, but that can be deceptive. Why not ask other people what they would think of as the opposite of 'aggressive'? When I tried this with members of an adult education class, they gave all sorts of answers, including 'kind', 'peaceful', 'quiescent', 'friendly' and many more. We ended up with about sixteen different words. You can see from these opposites that the word 'aggressive' means different things to different people. Our personal constructs are exactly that – personal and special to us. But they are an important part of the way that we represent other people to ourselves.

31
EVERYBODY KNOWS THAT

We are learning from the moment we are first born: our brains have evolved to do precisely that – it's why human beings have such a large brain relative to their body size. And we know that a lot of that learning involves developing mental representations. But we don't just take in information from our senses as if it was all essentially the same. We learn best and most deeply from other people. That's one of the reasons why schools involve real teachers rather than just machines, and why working with a friend is such a good way of revising for exams.

What we learn from other people isn't just factual. We also learn about our particular society and culture – not necessarily formally, that comes later. But we learn which actions are acceptable and which are not; how we should speak to different people; and also how we should behave in particular situations. Every society has its taboos – things which really shouldn't be done or even talked about because they are just not acceptable at all. There are the serious things, of course, like murder; but there are also social conventions. In Victorian society, for example, right across Europe it was taboo to talk openly about sex – that's one reason why the psychoanalyst Sigmund Freud found that his patients all had private, worrying obsessions about it. More recently, talking about your own impending death or dying has generally been discouraged in industrialized Western society, although this may be gradually changing.

In all societies, children are encouraged to speak in different ways to different people, depending on the social roles those people have. It is generally expected, for example, that elders will be treated with respect and spoken to politely. In urban societies, it is assumed that teachers or police will be addressed respectfully, while family members and friends can be spoken to much more informally. We have different kinds of **speech registers** that we use for this, including the stiff and formal one used by speechmakers or presenters; the consultative register we might adopt when asking a stranger for directions; the familiar language we use when talking with friends or colleagues; and the intimate language that we keep for very close relationships. We use these speech registers automatically: they are embedded in our social understanding as mental representations of how we should communicate in particular circumstances.

Speech registers: styles of talking which are appropriate in different social circumstances.

We also have distinct **social scripts** which prescribe how people should act on different occasions. Again, these are unconscious internal representations and we're not really even aware that we are using them. But we know, for example, how people should behave in a restaurant or in a cinema, and if someone strays from the script we are instantly aware that they are acting 'wrongly'. Whether we do something about it – well, that depends on our cultural and social representations too! I was recently in an airport lounge that gradually filled up with Scottish football fans travelling to their first open-audience match after over a year of lockdown rules: Scotland against England at Wembley. Suddenly there was a massive sound, as about a hundred excited fans all broke into song in unison. The nice thing about it, though, was that although it was clearly breaking the social script of how people usually behave in an airport, the airport staff were smiling and amused: they understood how important the match was to those fans and recognized that the singing wasn't doing any harm. So although

Social scripts: the socially acceptable patterns of behaviour that are appropriate to a given social situation.

Changing your ideas

Social representations have a consistent core, but they can be modified through conversation or discussion. You might, for instance, change your ideas about poverty after talking with a care worker; or adjust them in a different direction if you've spoken to an industrialist. Each of them will believe they know the 'real' answer, but are really expressing their own social representations. Your own core beliefs will tend to guide you towards one or the other type of explanation. If you feel that the conversation has given you more insight, you're likely to modify your social representation to encompass the new information. If you reject the new information, nothing will change. We learn from other people all the time, and our social representations are necessary cognitive structures which we use to make sense of our worlds.

such behaviour wouldn't normally be acceptable, and it was definitely breaking the social script, this time it was tolerated.

Social scripts and speech registers can be seen as types of schemas, cognitive representations which we use in our everyday actions. But we have other types of knowledge too – beliefs about the world, other countries, why life is like it is, and so on. These are social representations – forms of understanding that develop as a result of social interaction, and which are constantly negotiated through conversations and exchanges. They are the shared explanations and understanding about the world which underpin practically all aspects of our social and political understanding. Ideas about what constitutes the most effective form of government, about what's best for the economy, about whether people are naturally aggressive – these are all social representations that differ from one group of people to another and form the basis, not only of discussion, but also of political and social action.

32
TOUCH WOOD

With all of those different forms of representation available to us, perhaps it's not surprising that our thinking isn't always rational. Sometimes, what we do can seem distinctly odd to other people, even though it makes perfect sense to us. But then, as we've seen, we are drawing on a range of representations, including the schemas and concepts we've learned from childhood, our unique personal constructs and the social representations we have adapted from others to understand – and react to – what is going on around us.

The way we identify ourselves socially also makes a difference. Humans are social animals, and human society is organized into different groups. Most societies, for example, have separate groups for gender, age and occupations, and modern diverse society has many more. As we've discovered, we can classify just about everything, including BMW drivers, into groups, which means that we classify people too.

But how about the way we classify ourselves? How we see ourselves – our self-concept – is also part of the way that we think and act. You might, for example, see yourself as being reasonably current, well up on the latest trends. So if you were offered tickets for, say, a fashion show, you'd happily accept. Someone else, though, who saw themselves as uninterested in that sort of thing, might turn them down. Similarly, if you saw yourself as adventurous and physically active, you might jump at the chance of a sponsored desert trek, while someone who didn't see themselves that way might shudder at the very idea.

Social identification: the 'them-and-us' distinctions that allow us to see ourselves as 'belonging' to one social group or another.

We absorb our social classifications into our self-concept too. Not all of them all of the time, but just when they are relevant. A single individual might identify herself in various different ways, for example as a cyclist, a bookworm, a carer, a dog owner and a feminist. These are **social identifications**, and most of the time they won't actually be relevant. When they are, though, they can change, or even determine the way we think. For example, one day the person I described earlier in this paragraph – let's call her Jane – has cooked dinner, but doesn't really mind doing the washing up as well. She's about to start it when her male partner makes a sexist remark about kitchen work being a woman's job. Instantly, Jane's feminist identification comes into play. Now, the washing up is the last thing she is prepared to do. Until that point, she had one way of thinking about that task; after it, her thinking is totally different. It is shaped by her feminist social identification, which has suddenly become salient in that situation. That change of heart may appear irrational to an onlooker, and perhaps even to her sexist partner – after all, what has actually changed? But it isn't in the least bit irrational to Jane.

Superstitious pigeons

It turns out that even animals can develop superstitious behaviour. The behavioural psychologist B. F. Skinner investigated many aspects of training animals using rewards. He observed that it wasn't uncommon for a pigeon to develop a repetitive habit, performing a brief action like flicking its wing or nodding, before pecking the button that would give it a food pellet. Having done it once and then immediately received a reward, the bird had learned to connect the action with the reward, even though it was completely unrelated. Skinner also found that this superstitious learning was remarkably strong. Once it had been learned it was extremely difficult to train the pigeons not to do it.

Other things we do, though, can appear irrational because they really are. In chapter 12, we saw how important it is for people to feel that they are in control of things. But in some human societies, there's actually very little control that individuals can have over their lives. Farming communities, for example, are extremely dependent on the weather, as are fishing communities. Students often don't feel in control when it comes to exam time, and sportspeople may feel that their success is at least partly due to luck on competition day. As we learned earlier in the book, it's particularly stressful for people to feel out of control. So all of these groups tend to develop superstitions. They perform actions or beliefs which are really only **superstitious behaviour**, but they feel that these actions will bring them 'luck', or at least avert misfortune.

Superstitious behaviour: behaviour which is performed in the belief that it might influence a situation, when really it has no effect at all.

Many of these superstitions get handed down through families over time, even though they have little or no meaning in modern society. People touch wood when they've said something they hope will (or won't) come true; they avoid walking under ladders; they throw salt over their left shoulder when they spill it even though they don't believe in religion (the action was originally to blind the Devil who would be lurking there). It started as a way of trying to have a little control over a life that was essentially uncontrollable: at the very least, you could avoid bringing bad luck on yourself. Now, it's just a habit of thinking, of which we are often hardly aware.

6

REMEMBERING

Some memories last for years, while others last only a few seconds. But does that really matter? We retain information while we're actually using it and if it isn't important, we soon forget it again. What we do tend to remember are events and episodes in our lives, which are another type of memory. Even if we forget them, we can often bring them back to mind using hints or cues or recreating their contexts. And we rarely forget how to do things: the everyday skills we develop stay in our minds, even when our bodies become older and less capable.

We use language to remember things too. Memory for the meanings of words is an important part of communication. Special shared forms of language develop easily between friends, in families and professional groups. And the language that we use also affects what we can recall – sometimes even to the point of remembering things that didn't actually happen.

33
WHAT WAS THAT NUMBER AGAIN?

Psychologists have been studying memory since the end of the nineteenth century, and many of their early findings gave us insights that still hold true today. One of the pioneers of memory research, Hermann Ebbinghaus, studied it by memorizing lists of trigrams – groups of three letters that seemed like words but weren't, like VIL or KAD. Among other things, he discovered primacy and recency effects (finding we remember the first and last items in a list better than the middle ones), that even if we think we've forgotten everything we take less time to memorize a list on a second occasion, and that having two or three learning sessions is more effective than spending the same amount of time in one single session. We'll come back to his ideas in the next part of the book when we look at the topic of forgetting.

Other psychologists began to distinguish between short-term memory and long-term memory. Short-term memory is the sort you might use for a one-time security code to access an account. We only remember it until we've keyed it in – then we forget it straightaway. But other memories, like the memorable name we use for secure banking, are stored for much longer. They are in long-term memory.

One of the classic ways of measuring short-term memory

A popular misconception

The idea of memorizing through repetition has a long history in education, and is still commonly used by students who are revising for exams. Unfortunately, though, those students are actually adopting the worst possible method of revision. Yes, information can get into your memory using that technique, but it takes a very long time and isn't guaranteed to work. As you read through this section of the book, you'll be able to identify several much more effective methods of memorizing things. Simple repetition isn't nearly as useful as generating personal meanings, for example, or as linking new memories with what you already know. So if you or someone you know are trying to learn things by heart by simply repeating the material over and over again, try a different approach. You'll have far better results.

is the **digit span** test. Typically, people can remember between five and nine digits. Some can manage more or fewer than that, but one psychologist referred to it as 'the magical number seven, plus or minus two' because seven is the most typical. But we can remember more digits than that by chunking – that is, by grouping them together into meaningful chunks. If I asked you to repeat a sequence like 2-0-2-0-1-9-8-9-2-0-0-1, you might find it hard; but if you knew they were memorable years you'd see them as 2020-1989-2001, and that would be much easier to memorize.

Digit span: how many numbers or letters (collectively known as 'digits') someone can repeat from a list after hearing it only once.

Psychologists used to think that short-term memory was the first stage, and that information passed into long-term storage through being repeated. But now we know that it doesn't really happen that way. Rather, short-term memory forms an important part of what is now referred to as **working memory**.

Working memory is all about what we're paying attention to and thinking about at any one time. If you're trying to work something out or perform a complicated task, there are several elements involved. The working memory model suggests that we have a core central executive, which keeps the problem right at the forefront of our minds and does any working out that's needed. This receives information from various inputs, such as noises and sound representations, what someone is saying to us (which we process differently from other noises) and images or visual representations of what's involved. All of this information goes into an input register, which channels relevant information to the central executive. We also have what is known as an articulatory loop, which is like a kind of internal voice that goes over the information again and again, also feeding it into the central executive and helping to keep it on track. These elements all combine to give us an immediate working memory, which means that we can focus on a particular problem or activity using helpful information, but shutting out anything that's not important.

Working memory: a complex memory system that allows us to retain and use, temporarily, the important elements of a problem.

The idea of working memory is seen as more useful than short-term memory, mainly because it focuses on how we use our immediate memory rather than seeing it as a passive sort of process. Also, memories can last for different lengths of time. It isn't just a question of short-term memory existing for a few seconds and long-term memory being forever. Some memories stay with us for years, while others might only last for a couple of weeks and then vanish. The stuff you memorize for an exam, for example, often disappears from your memory once the exam is over and you don't need that information any more. But the memory of a single song can remain with you for the whole of your life.

34
WHAT DID YOU DO ON YOUR HOLIDAY?

Perhaps the main thing that we think about when we talk about memory is memory for the unique, personal experiences that we have had during our lifetimes. Our most precious memories are for the things that we did or the special moments we have enjoyed. Sometimes, we go over those memories time and time again – we never forget them because we have retrieved them so often. Other times, we may seem to have forgotten them, but they return to us as we're talking with an old friend or when we go somewhere which brings a memory back.

From the moment we are born – possibly even before – we are storing up personal experiences. How much we remember, though, is closely bound up with how often we bring those experiences to mind. A six-year-old often has quite clear memories of where they used to play when they were two – or at least, of the floor they used to play on. But those memories fade – ask the same question of a ten-year-old and they won't be able to recall anything about it. As we get older, we generally only remember things from when we were about four years old. We might remember a single, special episode from earlier, though.

Our personal memories – known as **autobiographical memory** – are hugely complex. How strong they are is dependent on two things: firstly, whether the memory has been retrieved since; and secondly, how special the event was. Someone who was asked to recall the playing floor they had when they were two, when they were seven, usually retains something of that memory for the rest of their life. Someone who fell into a river when they were a toddler and had to be hauled out might remember how that felt. But we don't remember routine events or situations very well. That isn't surprising: just think of all the things that can happen in one day. If we were able to remember everything, we'd be completely overwhelmed by all that information. So our brains organize it so that the memories we can retrieve most easily are those that actually mattered to us at the time.

Autobiographical memory: our personal memory for our individual life experiences.

That's partly because memories are so closely linked with our emotions. If something has gone well for someone we love, we feel happy for them. We enjoy spending time with friends or people who make us feel comfortable. If a friend is angry or impatient with us, we feel upset – in fact, we often get upset at such an experience even if it's someone we don't really know. The parts of our brains that deal with social interactions and relationships have strong connections with the parts that are concerned with emotions: they are not the same, but they are closely linked. So memories of our interactions with other people are automatically processed a bit more deeply than our other actions, and we remember them better.

You'd think, then, that memories which are linked with very strong emotions would hardly ever be forgotten. But that isn't the case. In one study, a researcher kept a meticulous diary over

The cognitive interview

Routine events just don't stick in the memory, no matter what age we are. Do you remember what you had for lunch the Tuesday before last? Chances are, you'd only remember it if something unusual happened on that day, or if your Tuesdays are somehow different from the other days of the week, particularly around lunchtime. It wouldn't come to mind as easily as, for instance, what you had for breakfast today. But that doesn't mean the memory is completely lost. See if you can reconstruct it. Think about that week and try to find cues to bring the memory back. What did you do on the other days of the week? Did anything distinctive happen on that particular Tuesday and, if so, was it before or after lunchtime? Where would you have been and what were the surroundings like? Were you with other people or eating alone? What would you have done after lunch? Forensic psychologists use cues like this to help people to reconstruct details of events that they think they've forgotten. They have developed a special form of interviewing, called the cognitive interview, which helps people to recall details of events. The richer the context, the more likely the memory is to come back. Try it – you'll be surprised at how effective it can be.

five years, noting down two significant events each day. Every few months, she chose two dates at random, read through the notes and tried to recall the events and when they had happened. Surprisingly, highly emotional events turned out to be no more memorable than any of the others, and some of them she simply didn't remember at all. So it's not just the emotion we feel at the time: it's also how it connects with everything else that matters.

Episodic memory: memory for specific events or experiences.

What we are talking about here comes into the general category of **episodic memory** – that is, memory for things or events. Episodic memory is quite different from the memory we use for how we do things: that's known as procedural memory, and we'll be looking at it more closely in chapter 36. Episodic memories can include knowledge we've gained from lectures, books or other people's stories, as well as things we've experienced directly. We may store or represent them in different ways: as images, muscle actions or even as mini videos that we replay in our heads. But essentially, they are all about things that have happened or experiences that we have had, and whether we recall them or not often depends on cues and contexts. We'll consider those in the next chapter.

35
WHERE AM I?

As we've seen, memories aren't always instantly available, but they can come back – sometimes a bit later than we'd like, and occasionally so clearly that they surprise us. Really, it's all about **memory cues**: pieces of information that connect with the stored memory and allow us to bring it to our conscious minds. One of the reasons why processing information enables us to remember it is because the processing we do helps to link that information with other cues and contexts. So the information doesn't stand on its own, but makes connections with other things that we know.

Memory cues: bits of information that are linked with other memories, allowing them to be retrieved.

Memory cues can take just about any form. Have you ever been transported back into the past by a song or a special piece of music? Or by a particular smell or taste, like the aroma of home baking or the flavour of peppermint? Sometimes we have memories that are so closely linked to sensory experiences like this that the minute we experience them, those memories come flooding back. Even a particular colour or shape might be enough to trigger some recollections if they were linked closely enough to the memory when it was first stored.

It's not only about cues, it's also about **contexts**. Our memories are powerfully connected to events and places – where we go and what we do is personal information that would have directly helped us to survive in our primitive past. So going somewhere you haven't been for some time can bring back a set of memories linked with that place.

Context: the general setting or situation in which something happens.

That's all to do with our being active in the world. We evolved as active animals and, as we saw in part 4, our perceptual mechanisms all help us to make sense of what is around us as we move through our lives. The same applies to memory. Our memories aren't just dry facts: they contain impressions of the places and social contexts in which we encountered them. Even if we are trying to recall some item of factual knowledge, the memory may be linked with an impression of the book or TV programme in which we first came across it. Being active allows us to experience a range of contexts, and that helps us to better remember what we come across.

If being active is so important, what happens when we aren't? Well, the answer is that we almost always have trouble with our memory – at least, regarding the things that are going on around us. People who are bedridden, or who have to spend their days sitting in one place, find that one day merges into the next and

The method of loci

The importance of the contexts and locations of memories is shown by the way we can use a physical location as a mnemonic to help us to recall information. It's an old method, dating back to records from the ancient Greek poet Simonides. Simonides had been delivering a speech at an indoor banquet and was called outside afterwards. An earthquake suddenly hit and everyone inside the building was crushed. The damage was so severe that the bodies were unrecognizable, but Simonides managed to identify everyone by recollecting details about the tables, remembering who had been sitting where. That's a dramatic example, but the method of loci ('loci' refers to 'location') is also useful for other things, like recalling shopping lists by visualizing where each item will be on the supermarket shelves.

it's particularly hard to remember what occurred when. It's as if our day-to-day activities act as a kind of personal, biographical index for our memories. If our lives don't change, through inactivity, the index is harder to maintain.

You may have encountered this when visiting an elderly relative who has moved into a care home. They often have perfectly clear memories of things that took place in the past, but they are much vaguer about what has happened recently. They may remember certain events, but can't quite pin down exactly when they occurred.

We don't find this unusual because we tend to believe that memory loss comes automatically with ageing. But that's another common myth. There's dementia, of course, which does affect memory. But dementia isn't all that common: it affects less than 10 per cent of old people in Western Europe, for example, and isn't usually anything to do with the memory loss experienced by elderly folk in residential care or those who have become passive or bedridden. Older people who are active and mobile don't experience anything like the same loss of memory. And there's research evidence that shows that younger people actually experience more absent-minded episodes than those who are actively retired. But older people tend to notice absent-mindedness each time it happens and worry that it's a sign of getting old, while younger people don't particularly notice those moments. How often have you walked into a room and forgotten what you went in there for?

36

WHAT AM I DOING?

When people experience amnesia, whether caused by an accident or due to some kind of neurological problem, they can find themselves unable to remember quite a lot of information. If it's very severe, they might even be unable to recall their own name. But there's one kind of memory which isn't affected, and that's remembering how to do things: the physical skills that we master when we are very young. Even with severe amnesia, people are still capable of actions like walking downstairs, opening doors, washing, shaking hands, talking, even riding a bicycle. All the little physical things that we do in our day-to-day lives.

These actions are known as **procedural memory**, and they are so commonplace that we barely think of them as being part of memory. But nonetheless, they are things that we have learned and have to remember how to do. Procedural memory can be interfered with by brain damage in some cases, but that's a very different type from the brain damage which causes amnesia. It's more likely to have affected the parts of the brain that control movement or muscle feedback than those parts of the brain which affect the memory.

Procedural memory: memory for how to go about doing things.

A lot of what we do is actually directed by procedural memory rather than active, conscious thinking. Our memories for some things have become so practised that they happen completely unconsciously – like remembering how to make a cup of coffee, how to tie a knot or how to kick a ball. They are represented differently in the mind – generally as enactive representation (see chapter 29) – and they are actually controlled by a

Still got it

Sports reporting in the media makes use of many retired sporting professionals as pundits, to comment on and discuss aspects of the action. It's helpful in lots of ways – partly because they have a good understanding of the rules and conventions of the sport concerned, and partly because they have often followed the careers of younger sportspeople and so can comment knowledgeably on what those people are capable of and what they have already achieved. The main value of the sporting professional as a pundit, however, comes from the skills that they already possess. In chapter 29, we saw how we use muscle memory to represent some of our knowledge. Our minds envisage what it is like to feel ourselves doing things, and that aspect of our thinking is particularly well developed when it comes to acquired skills. Retired Olympians or footballers still have those skills, mentally, which means that they can talk knowledgeably about what is involved, and understand the efforts that the current sportspeople are making, even if they can't perform at the top level themselves any more.

different part of the brain to that which is responsible for our conscious, deliberate actions. This is because they have become skilled actions – things that we are able to do accurately and without thinking.

There's a big difference between a conscious action and a skilled action. You'll have experienced this at first hand if you've learned to drive. At first, you have to think about every little activity that's involved – co-ordinating hand and foot movements, remembering how to change gear and when you should do it, checking the mirrors for other vehicles, and a host of other things. It's all very confusing and takes a lot of concentration. With enough practice, though, these actions begin to group together into smoother units and they don't require quite so much thought. You begin to change gears automatically, for instance, without having to try to remember what to do.

This chunking and smoothing of groups of actions happens through practice. It's fundamental to **skill acquisition** and it happens as the control of what you are doing shifts from the conscious part of your brain – the cerebral cortex – to a different part, known as the cerebellum. The cerebellum controls skilled movement and balance, and it is entirely separate from the cerebrum, which is where thinking happens. It's older than the cerebrum, evolutionarily speaking, which isn't really surprising – animals would need to have been able to move, run and react effectively long before having to think about the ins and outs of a potential predator's behaviour – and it has direct links with all of the basic brain mechanisms that keep our bodies going.

Skill acquisition: how we develop well-rehearsed and polished sets of actions or thought processes.

As previously noted, skill acquisition occurs through practice. With enough practice, we can develop all sorts of complex skills – and our bodies remember them. Once you've learned how to ride a bike or how to skate on ice, you don't really forget it. You might lose physical fitness as you age, so you're nervous about doing it, but the procedural memory for that skill remains with you and isn't generally affected by other types of memory loss. Even if you're completely confused about the world and who you are, you can still probably make yourself a cup of tea.

37
IS HE CATFISHING?

Have you ever been happily chatting away and then realized that your listener has understood something quite different from what you were talking about? Words and conversation are at the heart of our social natures. Children acquire language without really trying – from hearing other people speaking around them; from being actively taught as we gently correct them or repeat a word more clearly so that the child can pick up on its detail; and, of course, from their exposure to extended vocabularies through reading and school. Acquiring spoken language, unless we have a serious disability, is largely effortless.

Conversation, though, is more than just exchanging information through words. It has patterns, intonations and assumptions which colour how the words are used and directly affect their meaning. What's really interesting in how children acquire language is the way that they learn the social interaction part of conversations first. Even before they have words, children babble and make the kinds of sounds relevant to the language they are hearing around them. They use the timings and turn-takings of conversations and the intonations as well. Sometimes they can sound as if they are having a whole conversation with you – although none of it involves a single real word.

Those aspects of language – known as paralanguage – are essential to our understanding of the words. Think of a conversation involving several people, when someone turns to another and asks, 'And what do you think?' That question might be a genuine request to know, but if it's delivered in a certain way – like having the emphasis on the 'you' – it might actually be conveying the sentiment: 'I'm not actually interested because your opinion doesn't matter, but I'm asking because I feel as though I should.' Which is an entirely different meaning. That process can happen with just about anything we say: it all depends on the relationship we have with the other person and the context of those words. And, of course, the way the question is interpreted by the listener depends on those things too.

So using language doesn't necessarily involve clear communication. And the words we use can also have different meanings. For each of us, a word represents an idea or a concept – and it might not be the same idea or concept as someone else's. Asking 'Is he **catfishing**?' about an active fisherman on holiday means one thing, but posing the same question in regard to someone's dubious social media activity means something entirely different.

Catfishing: in social media, the act of representing oneself falsely or using a fake personality in order to deceive other people.

A word makes an association in our minds: it's a form of mental representation. But what that representation is partly depends on our individual experience. We saw in chapter 30 how our personal construct system – our idiosyncratic way of understanding the world – can influence how we understand words like 'aggressive'. We all have our own personal use of words, and some people develop theirs to such an extent that they virtually have their own private language. That's known as **idioglossia**, and it comes from the way in which we use particular words for special meanings in a form that wouldn't be understood by other people. Twins, too, can often develop a private language strictly between themselves. But if language use was all personal and idiosyncratic, we wouldn't be able to communicate using words at all. Our common experience of media, books and schooling means that, for the most part, a given word in a language generates common representations.

Idioglossia: an individual form of language spoken only by one person or, at the most, a single pair of twins.

But this isn't always the case. We're all familiar with the ways that groups of professions use their own particular jargon to communicate with one another – possibly even to make sure that other people don't understand them! And shared languages happen informally too. The common experiences of work colleagues, groups of friends or families means that they often develop their own 'language', which carries shared meanings that only they know.

I was once talking with a Finnish engineer who had spent some time working in the oilfields of the Middle East. His

Are you diglossial?

Most language communities have more than one form of language itself. It might be an informal version, used in everyday, friendly interactions, and a formal one, or it might involve regional or cultural variations. The same person, for instance, might speak conventional standard English when dealing with professionals or people they didn't know, but chat in broad Yorkshire dialect or Jamaican patois with friends and family. The two versions of language involve both different vocabulary and different grammar. The ability to speak more than one form of language in this way is known as diglossia. Being multilingual has been shown to be a distinct asset in terms of learning and brain development: it is possible that being diglossial has a similar effect in the brain, although not as strongly.

working group was completely multicultural and multi-ethnic, so they used English as a common working language. But what my Finnish friend had particularly enjoyed was how they had ended up more or less inventing their own language, with special meanings and ways of saying things. It wasn't anything like 'pure' English: it was their own special language unique to their group, but the flexibility of the English language allowed them to innovate in ways that are not so easy in other languages. They all understood one another perfectly, but it posed a distinct challenge to any newcomers.

38
WHAT WILL YOU DO ON YOUR BIRTHDAY?

There are many things that make human beings distinctive, and one of them is the way that we can think about things that haven't happened yet. Usually, if we think about memory, we tend to assume that it's all about remembering factual information or events that have happened in the past. But we also have a special kind of memory that we use for plans and intentions. It's known as **prospective memory**, and it comes into play when, for example, you remember that you planned to meet a friend to go shopping or that you have an appointment to get a haircut.

Prospective memory: memory for future things, like plans or events that haven't yet happened.

Of course, we don't always do what we plan to do – sometimes we just forget. That might be for emotional reasons, and we'll look at that in chapter 43. But it's more likely to be because we got distracted and just didn't remember in time. Researchers have identified five distinct stages in prospective memory, and it's a failure of the second one that's usually responsible for that type of forgetting. The five stages are:

1. Forming the intention or resolution in the first place.
2. A retention interval – the time that elapses between forming the intention and the occasion when it actually needs to be carried out.
3. Detecting the relevant cues which remind us that we planned to do something.
4. Retrieving the actual plan and what it involves from memory.
5. Carrying out the actions involved in the plan.

That second stage is the really tricky one. We need to go on with our everyday lives, with all the cognitive complexity that involves, while still, somehow, maintaining a kind of monitoring – for example, whether it is yet time for the thing to happen (I'm supposed to meet Jonathan on Friday, is it Friday yet?), or whether we are in the appropriate situation for that particular intention (I meant to ask Emma next time I saw her.

Did you forget what you came in for?

We've all experienced failures in prospective memory from time to time. They might be as simple as going into a room and forgetting the reason why, or forgetting to pick up some milk when we go to the shop. But sometimes they are more serious. Plane crashes may seem to happen often, but that's because they almost always hit the headlines. In reality, they are extremely uncommon, especially when you think about how many aircraft are normally in the skies at any one time. And when they do occur, there are intensive investigations into how they might have happened. What those investigations show is that these accidents are most often the outcome of failures in prospective memory. The pilot or crew may, for example, have been interrupted while going through their checklist and miss something out when they resume it; or an air-traffic controller may forget to clear a plane waiting on a runway to take off and then allow another plane to land there. Analysing how these prospective memory failures happen is useful because it helps to develop new warning systems or other ways of preventing such accidents from happening in the future.

Is she here?). Sometimes, though, there's so much going on that our unconscious monitoring fails. We get distracted and simply forget.

Have you ever had the experience of knowing there was something you ought to do, but not quite being able to remember what it was? That's because you've got as far as stage 3, but haven't yet managed stage 4. Actually remembering to do something also has an element of remembering the past – known as **retrospective memory**. We don't just need to remember *when* we intend to do something: when the time comes, we also have to recall *what* we originally planned: why we planned it and what we thought it would involve. Why, for instance, did you plan to meet Susan at the craft market next Tuesday? Why there and not somewhere else? There will have been reasons why that was a suitable time and a suitable place, and we need to recollect those reasons in order to act effectively. These are different cognitive activities, and researchers have found that slightly different areas of the brain are involved in maintaining intentions over time, and actually remembering what is intended.

Retrospective memory: memory for events and experiences that have happened in the past.

The final phase of prospective memory, of course, is fairly automatic: we're in the right place at the right time and we know what it is we need to do. But we can see that remembering to do things is quite a complicated cognitive activity. It involves intentions and planning, retrospective memory, monitoring time and other relevant time cues, and then, of course, doing whatever it was we intended to do. It's no wonder that it seems to be a distinctly human activity. Pets have intentions, as any pet owner will tell you, but monitoring them over time and avoiding being distracted by other situations is quite another matter. Although, thinking about it, distracting my pet spaniel from her intention of persuading me that it's her dinner time isn't actually all that easy!

39

DID THAT REALLY HAPPEN?

It's a common saying in the police force that if you have six witnesses to an event, you have seven different events. That's because although we all think our memories are accurate, we're actually terribly bad at remembering exactly what we have seen.

We've already found how our thinking is influenced by our beliefs and expectations, and they also affect what we remember seeing. But other things affect our memories too. This whole area of psychology was opened up by the work of Elizabeth Loftus, who conducted a study that has now become a classic. She showed a group of people a short film involving a car accident and then questioned them about it. Half of them were asked: 'How fast were the cars going when they *hit* one another?'

The rest were asked: 'How fast were the cars going when they *smashed into* one another?' A week or so later, the members of the groups were asked about the film again; in particular, whether the collision had resulted in any broken glass. The people who had been questioned about the cars hitting one another said, quite correctly, that there hadn't been any. But those who had been asked whether the cars had smashed into one another remembered seeing broken glass strewn across the road after the accident. They were quite insistent about it – it was firmly part of the incident that they remembered.

It became a classic study partly because these findings were replicated over and over again, but partly because it demonstrated so clearly that our memories are not just factual recordings of events. It's important, because so much of our justice system relies on **eyewitness testimony**. But Loftus showed that eyewitness testimony is notoriously unreliable: not only do we tend to remember what we expect to see, but our memories are also easily affected by what happens later – such as the questions we are asked about it and the way that they are phrased.

Eyewitness testimony: a report given by someone who has seen an accident or event take place.

This poses a distinct challenge for police and others who need to ask questions about events to find out what actually

Imperfect recall

Some people have what is known as eidetic memory – that is, they have an astonishingly accurate memory for details. At the time of the Watergate scandal in the US in the 1970s, one of the White House aides, John Dean, gave detailed word-for-word accounts of significant conversations that had taken place in his presence. Dean was known among his colleagues for his accurate retention, so his account was regarded as really important. Sometime later, the secret tape recordings that President Nixon's people had made of those dialogues came to light, and it became possible to compare Dean's versions with the actual verbal exchanges. What emerged had two sides. Dean's testimony was substantially accurate, in that he had remembered the social meanings of those conversations perfectly – what people had meant, and what their words had actually implied. But his memory of the specific words which had been used varied in many small ways. So, although he thought he had provided an accurate word-for-word account, he had actually given a true account of the participants' communications while getting a lot of the actual words and phrases wrong.

happened. In chapter 28, we saw how people try to make sense out of what they have seen or heard, and how that can involve adjusting the information to fit into our schemas. So that's one problem. But Loftus' research showed that even asking about it can distort a memory if the wording isn't phrased very carefully. The questions need to avoid even suggesting an answer. The **cognitive interview** that we looked at in chapter 34 was developed as a way of assisting investigators in getting round that problem, by helping people to recall their own memories without prompting them about the event itself.

Cognitive interview: a form of interviewing developed by psychologists that is designed to provide cues for accurate memory retrieval.

Before this finding became widely known, some police forces were even experimenting with using hypnosis to help people remember. But what hypnosis actually does is to put people into a very suggestible state, where they pick up on the tiniest cues and say or do what they think the hypnotist wants them to say or do. So, using hypnosis makes it particularly likely that someone's memory will be changed to what they think the questioner wants them to remember – and once a false memory has been implanted, it's impossible to tell it apart from a 'real' one as they're all completely real in that person's mind.

It's very difficult for us to accept that our memories aren't always right. Mostly, we can't go back and check, so there's no way of knowing. But here's one: have you ever gone to see a film that was a favourite of yours many years ago and then found that some scenes were different from how you remembered them? It wasn't the movie that had changed: it was your memory of it.

FORGETTING

Forgetting can take many forms and it happens for various different reasons. Sometimes we simply don't have the reminders that we need to bring a memory back; sometimes we can't gain access to the right memory because we're getting interference from something else; and sometimes we forget things because, deep down, we simply don't want to remember them. Then there is also the more dramatic type of forgetting or, more accurately, memory loss, which happens as a result of brain injury or disease. Studies of those living with these issues, as well as studies of people with remarkable memories, have revealed quite a lot about how memory works in the brain. Brain injury apart, though, Sigmund Freud believed that we only really forget things because in our unconscious minds we want to. It's possible that it could be an explanation for forgetting in exams, but we'll also look at how processing information makes all the difference to effective revision.

40

IS IT ON THE TIP OF YOUR TONGUE?

Do you do quizzes? Or watch quiz shows on TV? If you do, you're bound to have experienced those times when you definitely know the answer, but just can't bring it immediately to mind. It can happen in conversations too – when you can't quite recall a name, but you're certain you know it. You might even know what its first letter was, but the name itself eludes you. That's known as the tip-of-the-tongue phenomenon, or TOT for short, and it's a really common experience. It's your brain trying to access information that it has stored, but not quite managing to get to it at that time – although it almost always comes to you later on. So it hasn't really been forgotten: it just hasn't been retrieved exactly when you wanted it.

What you actually have retrieved is the **lemma** – the idea of the word as it's stored in your memory, before it has been converted into the appropriate sound and form of the word itself. As we saw in the last chapter, our memory really works on the meaning of what we've seen or heard, rather than the specific details, and the parts of the brain that store ideas and memories are different from that which deals with language. Both of those parts of the brain are involved when we try to remember the names of things or people, but they don't always work in time with one another.

Lemma: an early, abstract representation of a word before it has become formal speech.

Psychologists have been researching memory for over a hundred years, so quite a lot is known about it now. But one of the earliest findings was the way that we have different levels of forgetting. Hermann Ebbinghaus, back in 1884, memorized lists of **trigrams** and tried remembering them at different intervals. He identified what we now refer to as the four Rs – recall, recognition, redintegration and relearning savings.

Recall: when we easily bring to mind what it is we want to remember. It doesn't involve forgetting at all. If we were listing it as a level of forgetting, it would be level zero.

Recognition: when we can't recall something, but we recognize it as soon as we hear or see it, and we know that it is exactly what we were trying to remember. You've almost certainly encountered this if you watch TV quiz shows: you're trying unsuccessfully to think of the answer, but as soon as the contestant says it you know that it's correct. You had forgotten it, but not very deeply.

Redintegration: when you can't remember something and don't even recognize it if you see it, but if someone asked to you reconstruct it – for example, asking you to put a set of words in an order you had learned previously – you'd be able to get it back into its original form. So, even though you might have forgotten the information itself, you still retain some kind of memory of it.

Relearning savings: when something has been completely forgotten and couldn't be redintegrated even if you were given all its components. But if you tried to learn that information again, you would find that it would take you less time to learn than something you had never come across before. There is a level of familiarity which helps the relearning process.

Trigram: a nonsense syllable, i.e. a set of three letters that can be pronounced but has no meaning.

We have to remember that Ebbinghaus was identifying the four Rs from trying to recall long lists of meaningless nonsense syllables: in real life, we wouldn't be likely to come across specificexamples of redintegration very often. And, as we've seen, the complex nature of human experience adds a lot of detail,

Remembering in exams

The methods Ebbinghaus used may have been a bit artificial, but there is still something we can learn from his findings. Modern exams, for example, tend to require a combination of recall and recognition: essay questions are definitely about recall and multiple-choice questions are about recognition. But if we're sitting in an exam and desperately trying to piece together fragments of remembered information to produce an essay plan or an answer to a longer question, we might find that the bits we remember seem to follow one another in a certain way – which would be an example of redintegration. And the message about relearning savings, of course, is that it's always worthwhile trying to learn information during a course because, even if you forget it, it will save you time if you're revising the same material.

colour and context to our memories, and that can make all the difference. Ebbinghaus was trying to study 'pure' memory, so he deliberately cut out anything that could be meaningful from the lists that he was trying to memorize. Real world memory, as we'll see, is a bit different.

41

IS THAT FRENCH OR SPANISH?

A friend of mine recently bought a second-hand car. It was a make he hadn't owned before and the handbrake worked differently: it was a pull-out lever under the right-hand dashboard instead of a pull-up lever on the left between the seats. For quite some time, whenever he parked his car, he would find himself feeling for the handbrake with his left hand instead of reaching out with his right.

That's known as interference, and it's another common reason why forgetting can happen. As we've already seen, we organize information in our minds and link new bits of information with other, similar things that we've come across. It's all part of the process of developing schemas, as we saw in chapter 28. But sometimes, those bits of information can interfere with one another – when we're trying to remember one thing, a different (but connected) thing comes to mind.

Interference happens a lot when we are learning new languages, but it can occur with just about anything where there are several similar actions, skills or ideas. You might be trying to remember which type of car a particular footballer drives and only be able to think of a Ferrari. You may know that isn't right, but it's all you can think of at the time. Later, you might recall that he actually drives a Lamborghini. Your knowledge of Ferraris as being fast, expensive cars has interfered with your memory of Lamborghinis as another example of fast, expensive cars.

Interference can manifest itself in two ways. Firstly, it can take the form of **proactive interference**, when what comes to mind is something you have known for some time, instead of another bit of information that you've come across more recently or which you are newly trying to memorize. What this means is that you forget the more recent information because you're recalling the more familiar information instead. Your established, familiar knowledge interferes with remembering new or recent information, making you more likely to forget it.

Proactive interference: when material that is already known interferes with the learning of new information.

Secondly, it may be the new information that you remember and the older stuff that you forget. Imagine that you have just been learning Welsh because you're aiming to take a job in Wales fairly soon and you think it will be an asset. But then a friend introduces you to their cousin, who is French. As you try to recall some of the French you learned in school, all that comes to mind are the Welsh words for what you want to say. That's known as **retroactive interference** because new knowledge is interfering with previous memories – information from back in time, as it were. So that can be another reason why you forget things.

But forgetting can happen for many other reasons too. We saw in chapter 35 how important context is for remembering things. We can forget things simply because we're in an entirely different place with entirely different people. You might be at work, for instance, and think 'I'll remember to do that when I get home'. But you don't. Instead, it only comes to mind again the next day, when you're back at work. Personally, I've found that the only way to deal with that is to form a mental image of myself walking through the door at home and immediately doing whatever it is that I'm trying to remember to do. Putting the reminder in an appropriate mental context like that makes me much more likely to recollect what it was I intended to do, at the time that I need to remember it.

Retroactive interference: when new material being learned interferes with the recall of previously learned information.

We have internal contexts as well as external ones. The physiological state we are in is also part of our memories, and it can make all the difference to whether we remember or forget something. Perhaps the most common example of this is when we've been drinking alcohol. We might plan to do something during a social evening with friends and then have completely forgotten about it the next day. The memory only comes back the next time we have a drink or two. The condition of being slightly inebriated has become an internal context for that information, and we only recall it when we re-create that context by putting ourselves in a similar condition.

That's known as state dependent memory and it doesn't just apply with drugs. As we'll see in the final part of this book, different times of day can also affect our physiological condition, and so can emotions like being angry or calm. So, our internal states can also be factors in why we forget things.

Time to think

Interference can be a real challenge for people who commonly use more than one language. Moreover, researchers have found that the confusion can actually be measured in the electrical activity of the brain. In one study, people were asked to name pictures in either Chinese or English. Since the pictures used were the same in both languages (although the sequences were changed) you'd have thought that people would be quicker at naming them when they saw them for a second time. But actually, they were slower – not just in saying it, but also in thinking it. The difference showed in ERP measures (that is, measures of the electrical response time) in the parts of the brain that were involved. It didn't matter whether Chinese or English came first: the fact that they had already named the picture using one language meant that the word in that language was what came into their minds first, so they had to ignore that and then search for the word in the second language.

42
WHO AM I?

Amnesia is a favourite subject in film and TV dramas. Someone wakes up one morning and hasn't a clue who they are, where they are or how they got there. Usually this has been caused by the nefarious intervention of some villain or other, and various protagonists compete to encourage or discourage the return of that character's memory. It may make a good premise for a drama, but it isn't anything like real life.

In the real world, people with amnesia just don't forget who they are. Our sense of self is deeply embedded and hard to shake. Even those with severe dementia tend to remember who they are, though they may think they are living in the past and don't recognize the people around them. Just like the unconvincing way that hypnotism is portrayed on the big screen, the Hollywood version of amnesia is very different from how things actually are.

That doesn't mean that amnesia doesn't exist. It is a very real problem for many people. It's all about loss of memory – either memory for what happened in the past (retrograde amnesia) or an inability to remember new information (anterograde amnesia).

Retrograde amnesia is most likely to happen as a result of a stroke, a brain tumour, brain disease, a head injury or chronic alcoholism. It doesn't tend to involve problems with semantic memories like language, or procedural memories such as how to get dressed or make toast. General memories like that tend to be unaffected. But it does involve being unable to remember episodic memories – that is, relating to things that have happened in the past. Retrograde amnesia usually means that we're most likely to have lost memories of recent events. As the memories gradually return (as they tend to do after a head injury or when a tumour has been removed) it's usually the older memories that come back first – possibly they are stronger because they've been recalled more often in the past. The more often we recall something, the stronger the memory – an important thing to bear in mind if you're revising for an exam.

Retrograde amnesia: loss of the ability to remember past events, situations or people.

Anterograde amnesia is a different story. People with anterograde amnesia can remember things in the past quite well – or as well as anyone else, anyway. But what they can't do is store new information, so their long-term memories tend to be from before the onset of the amnesia and they don't recall anything new. In his book *The Man Who Mistook His Wife for a Hat*, Oliver Sacks described the case of a man with anterograde amnesia who had suffered brain damage when in his early twenties, which had been several decades before. Every morning he was shocked when he looked in the mirror and saw an old man looking back at him: he could only remember himself as the young man he had been before the brain damage had occurred.

Anterograde amnesia: loss of the ability to store new memories.

Research into anterograde amnesia has told us a lot about how memory works in the brain. One of the most important insights came from the case of Henry Molaison, who underwent brain surgery to control his severe epilepsy. The operation removed part of the brain's temporal lobe, and included a structure called the hippocampus. Following the surgery his epilepsy did improve, but he also found himself entirely unable to retain any new memories. He had a normal digit span and short-term memory, but his memories vanished after a few seconds and didn't remain for the long term. He could still learn new skills, though, but didn't remember doing the learning at all. Molaison donated his brain to science after his death, and from this and other cases, it emerged that the hippocampus plays an essential role in encoding and storing new memories.

Sadly, one of the more common causes of amnesia is a disorder known as Korsakoff's syndrome. It comes from long-term heavy alcohol consumption coupled with not eating regularly or adequately. This causes a lasting thiamine deficiency which produces neuronal damage: important parts of the brain degenerate, including parts of the hippocampus and the thalamus, and this results in a combination of both anterograde and retrograde amnesia. The anterograde amnesia means that the person has difficulty storing new memories, and that partly accounts for the retrograde amnesia too because they don't remember anything of the past few years.

People with this syndrome may be perfectly sociable and able to have general conversations: their disorder only becomes apparent when they are asked, for example, who the prime minister is, or about something specific that happened recently.

Acquiring the Knowledge

We know that damage to the hippocampus can cause problems with memory. But researchers have also found that exercising memories can cause the hippocampus to become larger than normal. In a classic research project, Woollett and Maguire studied London taxi drivers, who need to have the Knowledge – that is, to have learned every single one of the streets and byways of London. They compared the taxi drivers with London bus drivers, who spent the same length of time driving and experienced just as much traffic stress, but who only used established routes. MRI scans showed that, in comparison with the bus drivers, the taxi drivers had more neurones in those parts of the hippocampus concerned with storing spatial memories. But the other areas of the brain involved in their work, where the two groups had similar experiences, remained much the same.

43
DID YOU FORGET ABOUT THE DENTIST?

In chapter 38, we looked at our memory for events or things that haven't happened yet – prospective memory. But, of course, we don't always remember what we planned to do. Sometimes, we forget that we had arranged to meet someone or that we meant to go to the supermarket on the way home. That type of forgetting can happen through interference or because we don't have the right cues to remind us at the time. But it can also happen because we don't really want to remember it. We might forget that we have a dental appointment or that we said we'd make something for a bake sale because, unconsciously, we were dreading the dentist or because we really didn't want to meet the people who'd be at the sale.

It's known as **motivated forgetting**, and it's all to do with the way that our thinking can be influenced by our unconscious minds. It's tempting to see thinking as factual and rational, but as we've seen, it can be quite strongly influenced by our moods, emotions and associations. Sometimes we are aware of that: if we're angry with someone, for instance, we may realize – even at the time if we're very grown up – that we're only recalling the bad stuff about them, and that there have been other things they've done which were actually quite good. Most of the time, though, we don't realize until much later how our thinking was coloured by our emotions. Moods and emotions form an emotional context for our memories, so we're much more likely to remember things that fit with them and forget memories associated with a different emotion.

Motivated forgetting: forgetting which has served an unconscious purpose or wish-fulfilment.

We have unconscious wishes and desires as well, and these can also affect what we remember. This belief was popularized by Sigmund Freud, who developed the theory of **psychoanalysis** – a form of therapy based on the notion that human thinking is shaped and directed by the unconscious levels of our minds. It was a radical idea at the time Freud was putting it forward, because until then it had generally been assumed that the mind produced logical, rational thought, and that emotions and desires were quite separate experiences. Freud's theory proposed that our unconscious minds consist of a turmoil of demands coming from our emotions, our conscience and a great many (often painful) memories. The main job of the conscious mind, he argued, is to keep a realistic balance between all those demands.

Psychoanalysis: a form of therapy that involves identifying unconscious conflicts and interpreting them using a specific theory.

Freud and repression

Freud's central idea was that repressed sexual energies produced unconscious pressures on the mind – something that was quite shocking in Victorian society, which was quite prudish about sex. Perhaps for that reason, his theory became massively popular, taken up by literary figures, artists and intellectuals of the time. It is still used today by psychoanalysts. It's not much considered by modern psychologists, because we now know so much more about how the mind works. But Freud did open up the whole idea of the unconscious mind and how it can influence us, which was a challenge to the generally accepted opinion that the mind worked in conscious, rational ways. And some of his insights, like those of motivated forgetting and the idea of the unconscious mind itself, are important in our understanding of how we think, even if we don't necessarily accept the whole theory.

Freud saw all forgetting as being motivated forgetting in some way. We forget things, he argued, because they remind us in some way of buried emotional conflicts and early traumas. To bring them to the surface and become aware of them would be far too painful, so our minds suppress not only the conflicts themselves, but anything which could be remotely associated with them. We might forget about the bake sale because its organizer reminds us of a distressing encounter we had with someone who looked very similar; or we might forget about the dentist because the fear of someone poking about in the mouth brings back infantile traumas of having been punished for putting inappropriate things in our mouths.

44
MY MIND'S GONE BLANK

As you've learned, psychologists know quite a lot about thinking and memory. But what use is it? Well, one obvious time when we need to rely on our memories is when we're taking exams. Most of us have to do exams at some time in our lives, and often more than once. We might also need to do presentations at work, which involve telling other people about an idea or development. So how can all that research into memory and forgetting be put to good use?

Memories fade with time, but the more often we recall them, the stronger they become. That's why old people tend to remember their youth better than what's happened recently: those memories have been rehearsed much more often. Sometimes, though, it can take a few hours or days before a memory comes back to our conscious minds – and that can be disastrous in an exam. So revision is partly to learn the information to make it more familiar, but mostly to make sure you have enough cues to bring those memories back to your conscious mind when you need to.

How do we do that? It's all about **cognitive processing.** Some of the information that we receive comes to us passively, so we don't particularly think about it, and we forget it just as easily. Other things, though, are more meaningful. We think about them and connect them with other information that we know. In other words, we process the information and that means that we can recall it much better. And we remember best of all the things that concern our relationships with other people. That's part of our evolutionary heritage as social animals: social interaction and relationships are really meaningful to us, so we process them very deeply. This is why working with other people while you're revising can be even more effective than working alone – as long as you don't get distracted, that is.

Cognitive processing: mentally organizing and working through information so that it connects with other learned knowledge.

There have been a number of studies that have compared different levels of processing. In a key study by Fergus Craik and Robert Lockhart, participants were given long lists of words to learn. One group was asked to memorize the list by repeating the lists over and over again. A second group was asked to form a visual image of each word. And the third group was asked to make sentences using each word. The first group

Remembering lists

I once had a friend who was taking a medical course and needed to learn lots of sequences of medical terms and chemical names. She asked me if there was anything I could suggest to help her, and there was. The first thing was to convert each item of information into something she could visualize. So for each of the technical terms she created a mental image, usually based on the sound of the word. For example, the word 'agglutinogen' became a mental image of a glue tin. Then she used the method of loci that we explored in chapter 35 to visualize these images at different places along a familiar walk. A couple of weeks later, I asked her how she was getting on, and she said it was absolutely brilliant, although she was starting to run out of walks!

didn't remember much of the list at all – just the first and last few words. The second group recalled more of them, but the third group remembered almost twice as many as the second.

So we're much better at remembering meaningful information. **Visualization** – making a mental image of things, or changing their form in some way helps, but actually thinking about the meaning of the information you're trying to learn so that it makes sense to you, and connecting it with other knowledge that you have, is the best way of all. So revising in ways that make you think about the meaning of the information is going to be much more effective than any other approach. Summarizing that information in your own words, drawing diagrams to link one idea with another or drawing up sets of quiz questions to ask a friend who's revising for the same exam are all good ways, and I'm sure you can think of even more for yourself.

Visualization: forming a mental picture of an idea or object.

There's just one more important thing to know about exams and presentations: we don't keep all of our memories in our conscious minds all the time. We know lots of information, but it only comes to mind when we're reminded of it, through cues or questions. So, it's perfectly normal to wake up on the day of an important presentation or exam and feel that your mind is a blank and that you can't remember anything you've learned. Don't worry: of course you can't. If you've been asleep, your mind has been busy organizing and linking what you've learned with other things (more on this later), so that knowledge has gone deeper into your mind. Trying to keep it buzzing round your conscious mind all the time is pointless: you won't be able to retain all of the information. Instead, your mind will just make some elements seem more important than the rest. If you really understand what you've been learning, you won't have any trouble remembering it when you need it.

8

CONSCIOUS AND UNCONSCIOUS THINKING

The human mind is complex and multilayered: things can be just below the surface of our awareness or they can be so deeply buried that we are simply unaware of them – but they still influence how we think. We range from being brightly alert to 'switched off', daydreaming or dreaming. And even when awake, different times of day or night also affect how alert we are – as people who've experienced jet lag know only too well.

Our unconscious minds are much more influential than we might realize. We use defense mechanisms to protect us from threats, but we also have a powerful tendency to empathize with other people – one of the many positive, aspects of being human.

45
RIGHT HERE, RIGHT NOW

What exactly are we doing when we're thinking? Sometimes we're aware of our thoughts – we experience them as a kind of talking or pictures in our heads. A lot of the time, though, our thinking is in the background: we're barely aware that it's happening, although we can bring it to our immediate consciousness straightaway if we want to. But some of our thoughts, as we've seen, are so deeply buried that we're completely unaware of them.

All of which raises questions about consciousness and what it is. It's one of the things that we all think we know, but which turns out to be really difficult to pin down, and different theorists have different ideas about it. For the neurologist Mark Solms, for example, consciousness and awareness are much the same thing, so he argues that pretty well anything that is aware of its surroundings and can respond to them according to that awareness is conscious. Other theorists take a different view, such as seeing consciousness as being linked only with a form of self-awareness that allows us to see ourselves objectively – to step outside our minds, as it were. And some regard it as having emerged as a result of social interactions, which involved skills such as imitation, deception and language.

Psychologists, philosophers and neurologists will continue to argue about what consciousness is until the cows come home. But for most people, consciousness is what we feel when we're awake, it's distinctive to humans, and it probably involves self-awareness and being able to see things from another person's point of view. The last, though, is really tricky: that skill, which we call **theory of mind** or TOM, only emerges when we're between three and four years old. Before that, children tend to assume that what they know is what other people know too – like the small child who covers her eyes and assumes that you can't see her because she can't see you. But you'd be hard-pressed to defend the idea that small children are not conscious – and I think that might show just how difficult consciousness is to define.

Theory of mind: the ability to understand that other people have ideas and knowledge that are different to one's own.

So let's leave defining consciousness to the philosophers and neurologists, and just assume that it mostly happens when we are awake and taking in our surroundings. Our conscious perception of ourselves changes from moment to moment: for example, you may be reading this and, now I've mentioned it, suddenly become acutely aware of how you are sitting or standing. Your conscious attention has shifted from these words to encompass **proprioceptive** sensations and other types of input. As you carry on reading, those sensations may or may not stay with you – that depends on your level of concentration and also your surroundings. If you're on a train, for instance, you also need to keep part of your awareness focused on monitoring where you are and which station you've reached, while the rest of it focuses on what you are reading.

Are you mindful?

Have you ever *really* looked at an orange? That was one of the first exercises in the general approach that eventually became known as 'mindfulness'. The person would be asked to hold an orange and experience it fully – without eating it. They would focus on its appearance, including its shape, colour and texture, on its smell, on the way that it felt in the hand, and so on. The idea was that the process of doing this would allow them to shut out all the other thoughts that were intruding on their awareness and concentrate their minds fully on just one thing. That exercise was used in Gestalt therapy in the 1970s, and those ideas were later combined with Buddhist meditation techniques to form the therapeutic approach known as Mindfulness. Mindfulness is all about bringing your full attention to the immediate present and shutting out everything else. Perhaps because of the mental rest it gives people, it's been shown to have both psychological and physical health benefits, especially by reducing stress and anxiety.

Proprioception: the sense that tells us about the positioning of our limbs and when they move.

So consciousness isn't just an either/or event: we can be partly conscious of some things even while our main attention is elsewhere; we can be deeply immersed in something, but then suddenly switch our attention to our physical surroundings. We saw in chapter 21 how the sound of our own name can attract our attention even if we're concentrating on something else. Sometimes, for example if we're just waking up or if we're in a doze, we can be only moderately alert. And when we're 'switched off' and relaxing, we may or may not have conscious thoughts. As old Mrs Amelia Skinner said to Anne Shirley in L. M. Montgomery's *Anne of the Island*, 'Part of the time I sits and thinks and the rest I jest sits.'

46
DID YOU DREAM LAST NIGHT?

Even when we're not aware of our immediate surroundings, our brains are still active. There's a myth that we only use 10 per cent of our brains at any one time – that simply isn't true. The whole brain is active all of the time, but there are various pathways between groups of **neurones** that are involved with different types of mental or physical activity. Our brains are even active when we're asleep: while we are dreaming, of course, and also in the type of deeper sleep that we think doesn't usually involve dreams. (I say 'we think' because it used to be believed that dreaming only happens during REM sleep, but research has shown that we can sometimes dream during other types of sleep as well.)

Neurones: nerve cells that carry electrical messages between areas of the brain and the nervous system.

Dreaming in itself is an interesting phenomenon. Many theories have been put forward to explain why we dream. Some theorists have argued that it is simply random neural activity – our brain cells are continually active and our dreams are simply a by-product of that. Freud believed that dreaming was another example of unconscious conflicts and desires, almost always sexual in nature. These, he believed, became manifest in our dreams in symbolic form because facing up to them directly would be too emotionally traumatic. So for Freud, a tall tower in a dream would be a phallic symbol, while a train going into a tunnel would symbolize sexual intercourse. Not many modern psychologists would accept that view, but it has helped to sell a great many popular books claiming to have the key to interpreting dreams.

Realistically, though, everyone's experiences are different and it would be simply naive to think that everyone would interpret objects in the same way. What we do know, though, is that sleeping on a problem often helps us to see it more clearly in the morning. It's as if our unconscious minds have been working on it during the night and found a solution, or a way of dealing with it, that we didn't think of before we went to sleep.

There's a reason for that and it's all to do with **consolidation**. Throughout our waking day, we are bombarded with information of one sort or another and it doesn't stop coming even when we slow down. We have conversations, do work, relax, enjoy hobbies, watch TV, cook, eat, etc. – and all of these actions are providing cognitive input of one sort or another. We might shut some of it off for a while during a daydream or meditation session, but for the most part that input comes in continually. So, before we can really deal with one bit of information, even more is arriving.

Consolidation: the process of converting short-term experiences or memories into longer-lasting, meaningful ones.

During the night, though, we are much less aware of external stimuli and this makes it much easier for the brain to deal with the information it has been receiving. It can sort things into categories, link ideas together and connect the day's experiences with others we've had in the past. So it's not really surprising that it can find new solutions to things

What if you don't dream?

Dreaming helps our minds to sort things out and get them into perspective. For that reason, dreaming is important for our mental health, as well as our ability to think clearly. There have been several studies about what happens to people if they are prevented from dreaming (they are allowed to sleep but woken whenever they start to dream) and it's not nice. After a couple of nights, people become confused and very likely to make mistakes at work or college. As the dreamless period continues, they begin to have hallucinations, and often become quarrelsome and paranoid. Most of the time, these effects vanish after a good night's sleep. But if dream deprivation goes on for long enough, the person can become seriously mentally ill, ending up with what appears to be a permanent state of psychosis. Which is why it is really, really silly to get into competitions or challenges about how long you can stay awake. It's your own sanity you're risking.

that are bothering us: it's been able to connect aspects of that problem with other parts of our experience, to allow us to find alternative possibilities.

We usually have several dreams during the night, as our sleep pattern cycles between dreaming and non-dreaming sleep. Some people claim that they don't dream – they do really, but they habitually wake up from a non-dreaming sleep state, which means that they don't remember any of their dreams. Other people wake up from a dreaming state, so they can recall, at least for a few minutes, the dream they were just having. Even insomniacs dream, although they tend to dream that they are lying awake. One person I discussed this with said that on one occasion she felt as though she had been awake all night, and she only realized that she must have been sleeping when she checked the clock and found that two or three hours had gone by. So even if you think you can't sleep, there's value in resting quietly in bed.

Sometimes, we become aware that we are dreaming. These are known as lucid dreams. Psychologists investigating lucid dreaming have shown that people can actually make decisions and take some control over what happens in the dream. But that only works if what they decide to do is 'realistic' in that dream world. You can make someone enter the dream by coming through a door, for example, but not by just appearing out of nowhere. If you find you're having a lucid dream, try controlling it and see what you can do.

47

ARE YOU A NIGHT OWL?

Thinking, or at least conscious thinking, is closely linked with alertness. We think more clearly when we are focused on whatever we are thinking about. But our alertness isn't always the same: at certain times in the day we are fully awake and at other times we're a bit more dozy and relaxed. Our patterns of sleep cycle between deep sleep and lighter sleep, and our wakefulness goes in cycles too. These are known as diurnal rhythms or **circadian rhythms**. Diurnal relates to the daytime in the same way that nocturnal refers to the night, while circadian means around the day and night. So these are rhythms, or patterns, that we follow during the course of a typical day.

Circadian rhythms: biological and psychological processes that happen regularly around the clock – that is, at certain times during the twenty-four hours.

The most usual pattern is to come to wakefulness in the morning, maybe about 7 a.m., and to become gradually more alert until we reach a peak period at around 11 a.m. Following that there's a gradual dip, which can lead to a nap or at least a period of lower alertness round about 2 p.m., followed by a gentle rise towards a smaller peak in the early evening. From 8 or 9 p.m. our alertness gradually declines until we sleep. If we don't go to sleep that night, our alertness continues to drop, hitting an all-time low between 2 a.m. and 4 a.m., before steadily rising again towards 7 a.m.

That's a typical pattern, but it doesn't vary very much. Some people feel they are at their most alert during the daytime, so

Motorway accidents

Circadian rhythms affect us more than we realize. Traffic researchers have repeatedly found that the most likely time for motorway accidents is between 2 a.m. and 4 a.m., and that those accidents are mostly caused by drivers falling asleep at the wheel. We know that's the reason, because they are characterized by a total lack of skid marks. In a normal motorway accident, skid marks on the road show how the driver was braking hard while trying to avoid the collision. But if the driver has dozed off, they are not aware of the imminent collision until it happens – maybe not even then, if they've gone at full speed into a concrete bridge. Sadly, these accidents are common because far too many people believe they can stay awake all night and drive continually without losing concentration or alertness.

in the evening they find it hard to do things that involve a lot of concentration or complex thinking. Others, notably parents of small children, find that the only time they are able to do that type of work is at night, when they are not being constantly distracted, so they manage to train their minds and bodies to adapt to that time frame. And some people are 'night owls': they don't handle mornings well, but are more able to deal with complex tasks in the afternoons and feel most sociable at night. Their cognitive rhythms follow the same pattern, though, and daytime is still the best time for them to do complex work, although perhaps not first thing in the morning.

How do we know this? Researchers have found these patterns even in people spending time in large underground caves, where physical aspects of day and night don't exist. The patterns show in a range of measures, including physical ones like body temperature or measures of reaction time, and mental ones like our capacity for sustained concentration and how likely we are to make mistakes. It's that last one that's perhaps the most important, because in modern society, where people are often dealing with complex machinery, making a simple error can sometimes have dramatic consequences.

It's not the same for everybody, of course. Employees who work regular fixed shifts, for instance, can adapt to being asleep during the day and awake at night. But in many workplaces people have frequent shift changes – for example, having to adapt to a different time frame each week. That can be hard to cope with physically, as there isn't time for the body to adjust fully to the new schedule, but it's been shown that industrial errors can be significantly reduced if the shifts rotate forwards round the clock rather than backwards. It's easier to move from 6 a.m.–2 p.m. shifts, then to 2–10 p.m. shifts and to 10 p.m.–6 a.m. shifts than it is to switch from 6 a.m.–2 p.m. shifts to 10 p.m.–6 a.m. shifts, and so on.

Jet lag: the experience of feeling woozy and unable to concentrate, which is brought about by moving rapidly from one time zone to another.

You may have experienced the effects of your circadian rhythms yourself if you've travelled to a country with a very different time zone. **Jet lag** is a feeling of lethargy and lack of cognitive focus which happens because our bodies need time to adjust to a new circadian rhythm. It can take as many as ten days for our body clocks to adapt to new times, so businesspeople who have to do a lot of travelling across time zones and are not staying very long often take steps to ensure that their circadian rhythms are not too badly disrupted. When the US president Lyndon B. Johnson was abroad, for example, he was famous for eating his regular meals at the same times as he would have done at home – for example, having breakfast in the late evening and lunch in the middle of the night – and arranging his meetings according to US time when he was most wakeful.

48

DID I REALLY DO THAT?

'That's my fourth sandwich – I must have been hungrier than I thought!' Sound familiar? Sometimes we surprise ourselves with what we do or say. We have our own ideas about what we are like and what we are likely to do, but those ideas are not always entirely accurate. Our sense of self and who we are is known as our self-concept. We came across it briefly in chapter 9, when we were looking at the way that other people see us; and we've also looked at self-efficacy – how effectively we think we can do things. But the self-concept isn't fixed and definite: it changes over time, and sometimes in ways that we'd rather it didn't.

That's partly because we are so very responsive to what other people think of us. Many studies have shown how the feedback we get from others affects how we think about ourselves. If it's negative, it can reduce our self-confidence so that we become shy or anxious. If it's positive, it can help us to become outgoing and sociable. Or if it seems that people see us as having a particular trait or ability, we can begin to see ourselves that way too. Comedians, for example, almost always say that they came into that profession because friends or other people at school or work found their sayings or actions amusing.

Self-concept is generally thought to have two parts: our **self-image**, which is what we think we are like; and our **self-esteem**, which is how worthwhile we think we are. The two are connected, of course, but they are not exactly the same. They both influence how we think: seeing yourself as a quiet, thoughtful person might mean that you believed you would do best at jobs which fitted that idea of yourself; seeing yourself as fashionable might lead you to develop a social media presence as an influencer; and seeing yourself as being not very good at things might mean that you avoid challenges that you could do perfectly well.

Self-image: the general idea we have about ourselves and what we are like.

There has always been concern about the psychological effects of social media. Studies show that the psychological impact of social networking often depends on whether someone's activities are self-focused or other-focused. Social networking activity which is mainly about presenting your own self-image positively, tends to have beneficial outcomes, but social networking that is mainly focused on the idealized images of other people is more harmful, partly because it leads to negative comparisons and feelings of inadequacy. Selfies, though, are generally an effective tool for self-definition and self-expression, in the way that they open up possibilities for being creative in developing one's own self-image.

Self-esteem: the general sense we have about how worthwhile or deserving of respect we are.

Self-esteem can be powerfully affected by other people's reactions. People who grow up with facial disfigurements or other disabilities, for instance, often have to go through quite a psychological 'journey' before they are able to develop a positive sense of themselves: they have to learn to value the feedback they receive from those who really know them and to ignore negative reactions from strangers. Similarly, it can be hard for teenagers to learn how to deal with negative responses via social media. If they have developed a secure sense of self during their childhood, as most do, they are usually resilient enough to cope. But for many, negative feedback from social media can be really

The problem of us and them

Some researchers have found that self-image can become a significant factor in stereotyping and prejudice. There are those who seem to feel the need to defend and support their own self-image by forming very strict ‘us’ and ‘them’ categories, and then expressing contempt for people they see as ‘the others’. These attitudes become entrenched and exaggerated by associating with others of similar views – sometimes even leading to real violence, as has happened with extreme right-wing social media groups. Fortunately, though, most people are rather more balanced, and are able to be reasonably confident about their own self-image without feeling that they need to put other people down to maintain it. And the people who leave those groups are often astounded (and disgusted) by how extreme their attitudes had become while they were members.

damaging to their self-esteem.

So, the value or otherwise of social media does depend on how we use it. Modern living is all about being in contact with other people, and social interactions have always affected our thinking about ourselves. Only the way that we go about some of our communications has changed. As a general rule, research shows that people who use Facebook most often tend to have lower self-esteem than those who use it less frequently. Some studies have shown that taking a Facebook holiday for a week or two can significantly improve a person's mental health. But people who use Facebook to reinforce their own good qualities and get encouraging feedback can make positive improvements to their self-esteem. The most important thing, really, is learning how to avoid, ignore and/or dismiss sources of negative information and focus on things that help us to feel good about ourselves.

49
IT WASN'T ME!

As we've seen throughout this book, there are many aspects of our thinking of which we're entirely unaware. Mostly, they are simply not in our immediate consciousness, but can be brought to mind if we need them. But sometimes, as we saw in chapter 43, they are so deeply buried that they influence us without our even knowing they are there. Our unconscious minds protect us from facing up to potentially threatening or traumatic ideas by using a range of **defence mechanisms.** Psychotherapists and clinical psychologists are familiar with these, because they often encounter them as they try to help people to deal with psychological problems.

Defence mechanisms are surprisingly common, helping us

Defence mechanism: a method used by our minds, unconsciously, to avoid facing up to difficult or challenging information.

to cope with the personal threats and challenges that we face in everyday life. These are some of the main defence mechanisms that we use:

Repression: Blocking unwelcome thoughts or memories so that they don't come into our conscious awareness, such as forgetting the contact details of someone who was associated with an embarrassing or traumatic experience.

Rationalization: Finding logical reasons or explanations to justify a mistake in order to avoid facing up to being wrong, like claiming that you thought it was better to allow someone to oversleep when you've actually forgotten to wake them up as promised.

Denial: Simply refusing to accept or acknowledge real facts that would be disturbing if faced directly. It's not uncommon, for example, for someone who has lost a much-loved partner or spouse to have some spells of denial, when they feel the person hasn't actually died but is just out of the house.

Displacement: Redirecting a reaction or emotional response to a different target, because directing it directly at the very person causing it would be too challenging. An example of this is someone who has been given a hard time by their boss at work, and comes home and kicks the cat or, more likely, gets cross with their partner.

Compartmentalization: Drawing up and maintaining strict boundaries between one part of one's life and other parts. It's often used by people who have emotionally challenging jobs, like some army personnel or those who work with severely injured animals.

Projection: Attributing one's own unacceptable ideas or feelings to someone else rather than facing them directly – for example, an extremely irritable and intolerant person who claims that their workmate is the irritable one and not themselves.

Reaction formation: Acting, thinking or talking in a way that is the direct opposite of one's own real but unconscious feelings. The classic example of this is homophobia, where a person avoids facing up to their own latent homosexual feelings and instead expresses intense hostility towards gay people in general.

Sublimation: One of the more positive forms of defence mechanism, which involves channelling unacceptable emotions or ideas into some kind of productive activity. A typical example would be avoiding dealing with the personal trauma from a failed relationship by becoming intensely involved in writing or art.

Intellectualization: Avoiding facing up to the emotional dimensions of an event or experience by looking only at its abstract or intellectual aspects. It's not uncommon for divorcees, for example, to give financial or time-management explanations for why their relationship failed, rather than confronting its emotional aspects.

Regression: Reacting to challenging events or experiences by adopting the behaviour or emotional patterns of childhood or infancy, such as bursting into tears when things get too difficult to cope with.

Some of these defence mechanisms are distinctly unpleasant or unfair to other people, often leading to prejudice and aggression. But as we can see, some can actually be positive: sublimation, for example, has been claimed to have inspired much of the best work in literature and art; and even denial can give someone undergoing the turmoil and emotional trauma of bereavement a little bit of mental rest. It wouldn't be a good idea for it to continue long term, of course, but for an hour or so it's not necessarily a bad thing.

There was a trend in the 1960s to assume that all defence mechanisms were negative, asserting that people would achieve a more positive, healthy outlook if all their defence mechanisms were broken down. This led to T-groups (now called training groups), encounter groups and a lot of amateur interventions that didn't really work. Therapists quickly realized that this goal was simply unrealistic. Sometimes, we just do need to protect our thinking from the more traumatic underpinnings of our minds.

Can you cope without your phone?

Another way we protect ourselves unconsciously is through comfort objects. We can use certain items, places or activities in much the same way as a small child may use a favourite toy: as a refuge when things get stressful and a comfort in upsetting circumstances. It's a form of regression but heavily disguised, and as adults we may be entirely unaware that this is what we are doing. That doesn't stop us reverting to familiar, comforting actions when things get tough, though. We might retreat into the kitchen and get involved with some familiar baking or cooking, retire to a garden shed or 'man cave', or even go shopping. But by far the most common 'attachment object' in modern living is the smartphone, which has been shown to be a major tool for providing comfort when coping with negative emotions. There is even a new word, nomophobia, which describes the anxiety and even distress that some people feel if they are separated from their smartphones.

50
CAN I FEEL YOUR PAIN?

Thinking is an internal mental experience, concerned with how we, personally, process information. And a lot of our thinking has to do with other people. We think about them, we make judgements about them and they influence our thinking in many different ways. That's because we are essentially social animals: as a species we've evolved in social groups and our brains have evolved to be specially able to deal with social information. Some researchers believe that it is the complex demands of social living that actually caused the human brain to develop as extensively as it did. That was certainly a factor, although it might not have been the only cause.

One of the key aspects of social living is **empathy** – that is, to be able to put oneself in another's shoes and understand what they must be experiencing. Empathy is all about sharing, or at least understanding, other people's experiences. It can be expressed in many ways: sometimes it results in direct comforting – even two-year-old children have been shown to react to other people's emotional distress by trying to comfort them, for example by cuddling up to them or bringing them a favourite toy. But it can also be shown in more subtle ways: it's not uncommon, for example, for colleagues to deliberately lighten the workload of someone coping with emotional problems, or for neighbours to take food to someone dealing with a bereavement.

Psychologists have identified two kinds of empathy. The first

Empathy: the ability to feel or understand what another person is experiencing from within that other person's own frame of reference.

is affective empathy, and this is all to do with emotions, like feeling sympathy or compassion for someone, or experiencing personal feelings of anxiety or distress in the face of someone else's suffering. The other kind of empathy is cognitive empathy, which involves being able to understand someone's point of view or mental state. Cognitive empathy can include seeing things from another person's perspective (and even using those insights to achieve personal goals) or identifying with fictional characters from books or movies. This type of empathy is less concerned with emotions and has more to do with thinking, but the two are closely linked and can feed into one another.

People vary widely in how empathic they are. Some people don't seem to have much empathy: even if they are well meaning, they are mostly unaware of how other people are feeling. Others, though, can be deeply empathic: highly sensitive to their fellow human beings and as responsive to their emotions as if they were feeling them personally. And in some ways, they are. What is really interesting is the way that empathy seems to be a basic feature of how our own brains work. We have special areas of the brain that are involved in communicating with others or interacting with them. Those areas also contain nerve cells, or neurones, which generate reactions in our own brains that reflect what we are witnessing in other individuals. They are known as **mirror neurones**. So seeing someone smile, for example, activates mirror neurones that make us feel inclined to smile too, or watching an athlete do a long jump causes a slight echo in the parts of our brain which control the muscles involved in jumping. It's another way that our social nature has affected our brain development, and it shows how deeply empathy is embedded in our thinking.

Mirror neurones: cells in the brain that control our own responses, but which also respond to seeing other people act in the same way.

Empathy is linked with the idea of emotional intelligence: the way that we use the emotional aspects of living to manage not just our personal thinking, but also our relations with others. Empathic people tend to score highly on measures of emotional intelligence and have been shown to make good leaders and managers, as well as being particularly competent in caring professions, such as nursing or social care. But empathy isn't just about personal or interpersonal traits: it has been described as the oil that lubricates social living, and it does that in all sorts of ways. Apart from individual acts of sympathy or kindness, people can express empathy on a mass scale. Studies of crowd behaviour in Olympic and Paralympic events have shown how spectators can become caught up in athletes' personal stories, demonstrating massive support even if that person isn't all that successful. And charity fundraising, of course, also draws on people's empathy by showing individuals or animals in need –

with huge success. Empathy is such a fundamental part of being human that it occurs in all sorts of ways, in all sorts of contexts.

How empathic are you?

Not everyone experiences empathy to the same degree. Psychologists have investigated what they call the Dark Triad, which is a set of personality traits that can be dangerous or undesirable at their extremes, although many people have them to a limited degree. The Dark Triad comprises three traits: narcissism, typified at its extremes by pride and egotism; psychopathy, characterized by impulsivity and selfishness; and Machiavellianism, which manifests itself in an exploitative attitude towards other people and an absence of morality. Although a lack of empathy is most strongly associated with narcissism, a high score in any of these three traits implies that the personality is pretty well lacking in empathy or warmth towards their fellow human beings. Fortunately, these extremes of personality are rare, and a lower level of those traits can sometimes be beneficial. A complete absence of narcissism, for example, would be someone with no self-belief at all.

INDEX

O

P

R

S

U

V

W

Z